T0215418

HYDRAULIC CONTROL SYSTEMS

Theory and Practice

HYDRAULIC CONTROL SYSTEMS

Theory and Practice

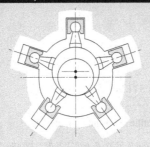

Shizurou Konami
Takao Nishiumi

National Defense Academy of Japan, Japan

World Scientific

NEW JERSEY · LONDON · SINGAPORE · BEIJING · SHANGHAI · HONG KONG · TAIPEI · CHENNAI · TOKYO

Published by

World Scientific Publishing Co. Pte. Ltd.

5 Toh Tuck Link, Singapore 596224

USA office: 27 Warren Street, Suite 401-402, Hackensack, NJ 07601

UK office: 57 Shelton Street, Covent Garden, London WC2H 9HE

British Library Cataloguing-in-Publication Data
A catalogue record for this book is available from the British Library.

HYDRAULIC CONTROL SYSTEMS
Theory and Practice

ISBN 978-981-4759-63-2
ISBN 978-981-4759-64-9 (pbk)

Typeset by Stallion Press
Email: enquiries@stallionpress.com

Printed in Singapore

Contents

Preface

The hydraulic system is a transmission device that uses oil as medium. It has the capability of transforming the mechanical power of a prime mover into fluid power, and then converting it into the mechanical power required at specified locations with flexibility. Since hydraulic control systems feature high power density, quick response and precise accuracy using digital technology, they have been widely used as energy transmission devices in aircraft, ships, construction machinery, machine tools, and others. Therefore, it is indispensable for a mechanical engineer to be well-versed with hydraulic control technology. Although the technology relates to a wide field of engineering, it is primarily associated with fluid mechanics and control theories. This book provides a comprehensive treatment of the design and analysis of hydraulic control systems, which will be invaluable for practicing engineers as well as undergraduate and graduate students specializing in mechanical engineering.

The book focuses on how to apply the basic theory to practical problems in the use of hydraulic components and systems. First, the fundamental concepts of hydraulic control systems are addressed, and illustrated by reference to applications in the field of aviation engineering. Next, the fluid mechanics necessary for the comprehension of hydraulic elements are presented. The technology of hydraulic components composing hydraulic control systems is addressed. Finally, fundamental control technology is discussed, together with its application to hydraulic servo systems. This includes the formation of hydraulic servo systems, basic control theorems, methods identifying the dynamic characteristics of hydraulic actuator systems, and a design method for hydraulic control systems. Exercises that

will help readers recapitulate the chapter contents are provided at the end
section of each chapter.

We deeply appreciate Dr. Robert Drayton in Cardiff, U.K. and Prof.
Andrzej Sobczyk in Cracow, Poland who proofread this draft and pointed
out the technical aspects and English grammar in advance. It should be
also noted that the original work was published in Japanese in 1999. We
would like to thank Tokyo Denki University Press for their permission to
use the original illustration data presented in the figures.

Shiurou Konami and Takao Nishiumi

Yokosuka, Japan

May 2016

Chapter 1

Introduction

In this chapter a simple hydraulic system is explained in detail, in order to clarify the principles of power control transmission making use of a liquid. Practical applications of hydraulic control systems are then illustrated using examples drawn from the field of aviation.

1.1 Basic Concepts and Formation of Hydraulic Systems

A hydraulic system is an energy transmission control system based on the use of a liquid. It transforms the mechanical energy of a prime mover into fluid energy, controls the energy in a fluid configuration and reconverts it into mechanical work at the chosen location. Since the fluid energy is both transmitted and controlled, power may be controlled by means of hydraulic valves driven by low electric power and low hydraulic power. When linked to a computer, the energy control transmission system offers several desirable features such as high power output, continuous and precise control, high-speed response, etc. Therefore, hydraulic control systems are widely applied in the transmission control systems in aircraft, ship, construction machinery, machine tools, etc. The equipment comprising hydraulic systems can be classified as follows.

(1) *Energy-transforming components*

There are various types of components that transform the mechanical energy of a prime mover into hydraulic energy. These are called hydraulic pumps or, more correctly, positive displacement pumps. An element that transforms hydraulic energy into mechanical work is called a hydraulic actuator, consisting of a hydraulic cylinder and a hydraulic motor. The hydraulic cylinder is an actuator performing a straight motion, whilst the motor is an actuator performing rotary work.

(2) *Components controlling hydraulic energy*

There are three kinds of control valves. These are pressure control valves to control pressure, directional control valves to control flow direction and flow control valves to control the flow rate in the hydraulic circuits. These are classified as components controlling the hydraulic energy. An accumulator accumulating fluid energy or absorbing the surge pressure may be classified as an energy-control element.

(3) *Energy transmission components*

Transmission lines, fittings, oil reservoirs, filters, heat exchangers and accumulators are classified as energy transmission components. They are also referred to as hydraulic system components.

A hydraulic pump is a positive displacement pump, which pushes out the fluid trapped in a pump chamber. Figure 1.1 illustrates the operating principle of a positive displacement pump. Each cylinder port has a check valve, one of which connects to the line with a reservoir while the other connects to the line with an actuator.

A check valve is a directional control valve which makes the fluid flow in a constant direction. As shown in Fig. 1.2, the flow in the direction indicated by the arrow passes through the check valve because a ball with a spring in the check valve is pushed aside by the differential pressure of the flow. However, flow in the opposite direction does not occur because the differential pressure on the ball pushes it back to close the flow passage.

In Fig. 1.1, the hydraulic fluid flows into the pump chamber through a check valve connecting a line with a reservoir, provided that the piston

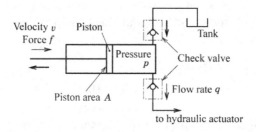

Fig. 1.1   Pump model.

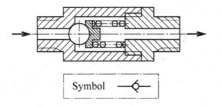

Fig. 1.2   Check valve.

is moved in the left direction by the prime mover. Similarly, the hydraulic fluid in the pump chamber flows out through a check valve, provided that the piston is moved to the right by the prime mover. Moving the piston with a reaction force f at velocity v, the power w is transmitted to the piston by the prime mover, as shown in the following equation:

$$w = fv. \tag{1.1}$$

Defining the displacement volume of the fluid per unit time as the flow rate, the flow rate q pushed out by the piston with the surface area A is described as follows:

$$q = Av. \tag{1.2}$$

Considering a pressure p that is generated in pump chamber depending on a load, the reaction force of piston f is denoted as follows:

$$f = Ap. \tag{1.3}$$

The equation of power w is derived from Eq. (1.1) to Eq. (1.3)

$$w = pq. \tag{1.4}$$

Equation (1.1) yields the mechanical power fv, and Eq. (1.4) yields the hydraulic power pq which is transformed by the piston of the pump. Then the power is transmitted through the transmission lines in the form of a fluid configuration with pressure p and the flow rate q.

The output power of the system per unit system mass is called power density. High power density is one of the features of hydraulic control systems. Recalling Eq. (1.2) to Eq. (1.4), the power density rises by raising the system pressure, because the system components with small bulk can be adapted to the required power ratings by using high pressure and small flow rates. This is why high pressurization in hydraulic systems has been widely promoted.

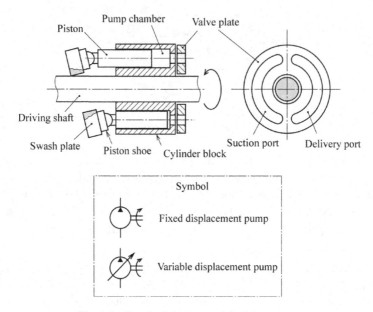

Fig. 1.3　Swash plate type axial piston pump.

Figure 1.3 illustrates a swash plate type axial piston pump that consists of a cylinder block connecting with a driving shaft, a swash plate with piston shoes (slippers) and a valve plate (port plate). The cylinder block has several cylindrical bores located at a constant radius, and a piston is put into the bore. Though the piston shoes slide along a swash plate that is fixed at a constant inclination, they are able to move along a circular path on the swash plate surface. Considering the rotation of a cylinder block with pistons inserted in bores, each piston reciprocates along with the rotation because slippers connecting to the end of piston move along a circle on the swash plate surface. Then, the other end of cylindrical bore faces the suction side of the valve plate in the region increasing the volume of pump chamber, and it faces the delivery side of the valve plate in the region where the volume of pump chamber decreases. Therefore, a swash plate type axial piston pump is able to discharge hydraulic fluid continuously, without any check valves.

The piston stroke may be controlled by varying the plate's inclination. Thus, a variable displacement pump can control the pump displacement by changing the inclination of the swash plate when the system is in operation. In contrast, a fixed displacement pump has the swash plate with a fixed

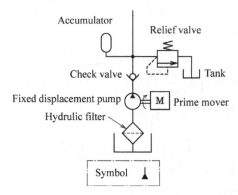

Fig. 1.4   Hydraulic source device.

inclined angle. Thus, it discharges the fluid at a fixed flow rate provided that the rotational velocity of a driving shaft is a constant.

Figure 1.4 shows a hydraulic source device called a tank unit using the standard graphic symbols for fluid power diagrams. Driving the prime mover, hydraulic fluid is suctioned into the pump chamber in a fixed displacement pump through a suction line from a reservoir, and then the pump discharges hydraulic fluid with fluidic power. The hydraulic fluid passes through a check valve and flows into a transmission line with an accumulator and a relief valve. The accumulator discharges fluid with fluidic energy when the pump displacement becomes temporarily insufficient. It has another role which absorbs the pressure fluctuation and surge pressure in the transmission line. The relief valve limits the maximum pressure in the transmission line.

Assuming that the fluid flow in a transmission line comes to a halt by closing the ports of the directional control valve, then, theoretically the pressure p could rise to infinity because fluidic power pq transmitted by the prime mover remains finite and the flow rate q becomes nearly zero. The relief valve and accumulator are provided in the transmission line, located near the pump, to avoid such sudden and excessive pressure changes. The relief valve maintains the pressure in the transmission line below the maximum permissible level by sending part of the pump flow rate through a tank port.

Figure 1.5 shows a functional structure of a balance piston type relief valve with a pilot valve having significantly small pressurized area compared to the main poppet valve.

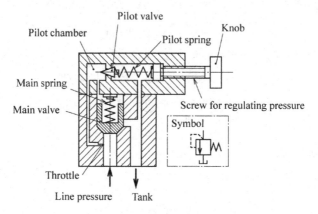

Fig. 1.5   Relief valve.

When the line pressure is lower than a cracking pressure corresponding to the pilot spring force, this force closes the pilot valve port. Then, the main spring force also closes the main valve port because the differential pressure between the upper and the lower does not occur. When the line pressure becomes higher than the cracking pressure, the pilot valve port begins to open. Then the differential pressure between the upper and the lower surface of the main valve is generated by the flow passing through a throttle in the pilot valve line, so that the main valve opens slightly and some of the fluid in the transmission line flows into the tank line through the main valves. Provided the deflection of the pilot spring is adjusted by operation of the knob with a spring so that all of the pump displacement passes through the main valve port at the allowable pressure in the transmission line, whose pressure keeps below the allowable maximum level.

As shown in Fig. 1.5, the standard graphic symbol of a relief valve consists of symbolic elements i.e. a broken line, an arrow and a block with a spring symbol and a reservoir symbol. The arrows in a block stand for the flow in a relief valve whilst a broken line denotes a pilot line pressure that controls the opening of the main valve port.

Figure 1.6 is an illustration of a diaphragm type accumulator. An elastic diaphragm separates the vessel of the accumulator into two parts. The upper chamber of the diaphragm is previously filled up with nitrogen gas that has about two-thirds of the system pressure, whereas the lower chamber of the diaphragm is connected to a transmission line of the hydraulic system.

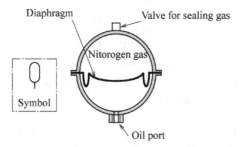

Fig. 1.6   Accumulator (diaphragm type).

As the hydraulic fluid in the transmission line flows into the lower chamber, the change of the gas volume in the upper chamber absorbs pressure ripples and surge pressure in the transmission line. It also performs the supplementary task of supplying the hydraulic fluid when the pump flow rate becomes temporarily insufficient. Thus, the hydraulic source device supplies the hydraulic fluid to the hydraulic control circuits so that it is transformed into mechanical work by the actuators at the desired locations.

Figure 1.7 illustrates a reciprocating function of an actuator as compared with diagrams using graphic symbols. Figure 1.7(a) shows a graphic symbol of a directional control valve (four-way three-position type) that controls the directions of flow. The ports A, B, P and T close in the central square block and the arrows in square blocks on both sides indicate the flow direction in the valve whose port position is alternated to each square block position. The black triangular symbols in Figs. 1.7(b)–(d) signify hydraulic source devices similar to Fig. 1.4. When the valve spool is located at the central position, as shown in Fig. 1.7(b), the pressure of the transmission line is kept below the allowable maximum value set by means of the pilot spring force of the relief valve. Then, the hydraulic fluid in the transmission line is drained through the relief valve. When the spool moves to the right or left, the hydraulic fluid in the transmission line flows in the directions indicated by the arrows in Figs. 1.7(c) or (d).

Figure 1.8 shows the flow characteristics of a relief valve. The pilot port of the relief valve opens at a cracking pressure and at that instant the flowing liquid passes through the pilot port. When the pressure in the transmission line becomes slightly lower than the allowable maximum pressure, the main valve port opens, and the flow rate through the main port then increases rapidly. The fluidic energy converts to thermal energy when the drain flow passes through the narrow passageway of the main valve port.

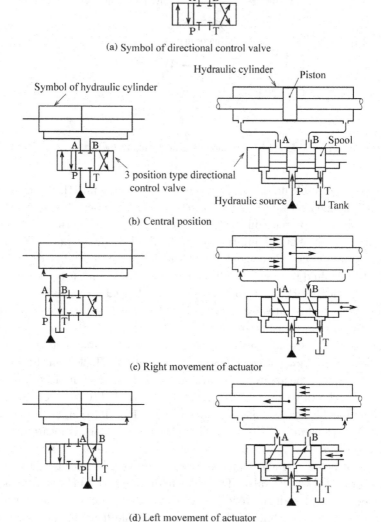

(a) Symbol of directional control valve

(b) Central position

(c) Right movement of actuator

(d) Left movement of actuator

Fig. 1.7   Reciprocating motion of actuator.

This leads to an increase of the temperature of hydraulic fluid instead of a transformation to beneficial work. Temperature rises at the expense of the energy and this can lead to serious problems if it continues for a long time.

The use of a center-bypass directional control valve is one of the methods used to prevent such an energy loss and associated problems. Figure 1.9 shows a graphic symbol of a center-bypass type four-way three-position

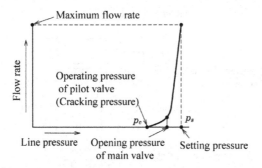

Fig. 1.8   Flow characteristics of a relief valve.

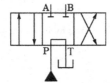

Fig. 1.9   Graphic symbol of center-bypass type four-way three-position directional control valve.

directional control valve. This valve passes directly through the pump flow rate to a tank at the central position of a spool where the spool closes the two connection ports A and B in an actuator. The other method uses the hydraulic source device with a pressure compensated variable displacement pump.

Figure 1.10 shows a hydraulic system using a hydraulic source device complete with a pressure compensated variable displacement pump and a directional control valve operated by solenoids. The main function of a pressure compensated variable displacement pump is to control automatically the inclination of a swash plate, in relation to the load pressure, so that the pump displacement becomes nearly zero at the allowable maximum pressure. Figure 1.11 shows the cut-off characteristics of a pressure compensated variable displacement pump. The curve of the cut-off characteristics reveals that the pump displacement decreases rapidly when the pump pressure exceeds the cut-off pressure. Therefore, the energy loss of the hydraulic system is significantly reduced by means of the cut-off characteristics.

The basic composition of a hydraulic system has been now clarified using simple examples. It is clear that, using the various hydraulic circuits, we can design control systems to meet most requirements of hydraulic system.

Hydraulic cylinder

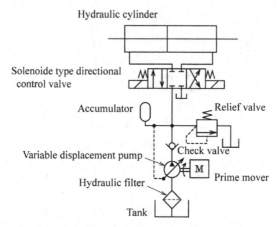

Solenoide type directional
control valve

Accumulator

Relief valve

Variable displacement pump

Check valve

Hydraulic filter

Prime mover

M

Tank

Fig. 1.10   Hydraulic control system using hydraulic source device with a pressure compensated variable displacement pump.

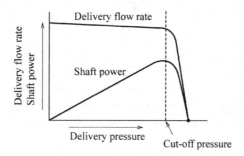

Fig. 1.11   Cut-off characteristics of pressure compensated variable displacement pump.

1.2 Overview of Hydraulic Control Systems

The key features of a hydraulic control system compared to the electric and mechanical control systems are as follows:

(1) High transmission power is controlled by means of hydraulic control valves driven at small power consumption. It may be controlled very precisely using a computer.

(2) A hydraulic control system features high power density, high-speed response and a sufficiently high efficiency provided that power transmitted exceeds about 1.0 kW.

By virtue of the above-mentioned features, hydraulic control systems have become indispensable in power transmission-control devices used in

1 Automatic sighting device in radar system
2 Tracking antenna control system
3 Missile attitude control system
4 Stability augmentation system in airplane(SAS)
5 Air flow rate control device in engine
6 Maneuvering device in airplane
7 ICBM thrust direction control system
8 Thrust direction control of space shuttle
9 Anti-skid brake in airplane

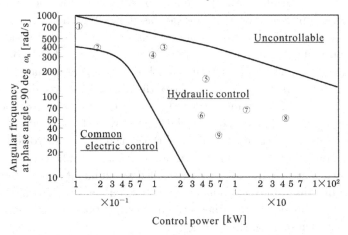

Fig. 1.12   Required control characteristics in aviation field.

aircraft, ships, construction machines, heavy vehicles and ocean development machinery, etc.

The control characteristics shown in Fig. 1.12 reveal that there is a controllable region in the transmission power versus response speed relationships, and that a hydraulic control system acts as a control device with high power and high-speed response. High-pressurization in hydraulic control systems has been promoted in order to improve the high power density and high-speed response. Currently, a pressure of 35 MPa is used in the hydraulic control system of construction vehicles. An aircraft hydraulic control system with a 56 MPa pressure system is being developed, though a 21 MPa pressure system has been effectively used for 40 years and more. It is anticipated that the development program will achieve the mass reduction by 30% and volume reduction by 40% in the aircraft hydraulic control system with an output power 75 kW or more.

The primary problem in extremely high-pressure hydraulic control systems is an increase of inner leakage. High pressurization requires that the clearance between a moving part and its surrounding wall should be reduced. This conflicts with the need for sufficient clearance to provide lubrication of the moving part. High pressurization has been achieved by means of improvements to lubrication characteristics following the development of new materials, improvements to hydraulic fluid, advanced processing of the material surface and so on.

Application of hydraulic control systems in the aviation field can create problems relating to the possibility of a sudden breakdown without manifesting any moderate deterioration as it is often the case with electronic equipment. For instance, the automatic flight control system and the stability augmentation system (SAS) fail to function if the nozzle of a servo valve in the hydraulic system is choked with a contaminant from the hydraulic fluid. The subsequent failure of the hydraulic source device would disable the system for maneuvering the aircraft. Thus, the breakdown of the hydraulic control system during the flight may lead to a serious accident because it steers the aircraft control surface such as the ailerons, rudders and elevators. To avoid such accident, an aircraft hydraulic control system should incorporate a multiplex system called a redundant system to handle any system failure. In a redundant system, even if a system line in the multiplex system breaks down, other system lines will take over automatically and the accident can be thus, avoided.

Figure 1.13 illustrates the hydraulic control system in the Boeing 767 passenger airplane. It is a triple redundant system. The two hydraulic system lines in the triple redundant system have pumps driven by different engines and also pumps driven by AC motors complete with dynamos. The other system line has a pump driven by an air pressure unit and a pump driven by a ram air turbine (RAT) in addition to the pumps driven by the AC motors with the dynamos.

One current type of aircraft steering system operates in a system mode that is called Fly-By-Wire (FBW). In a hydraulic control system of the FBW, the direct mechanical connection between the control stick and the aircraft control surface is removed. The control surface is controlled by means of hydraulic actuators with servo valves that are operated by the electric signal generated by the motion of the control stick.

Although the FBW system seems more affected by adverse factors such as a temperature, magnetic field and electric jamming than mechanical steering systems, it has the flexibility to utilize unused space, and contributes to reduction of weight. Moreover, the FBW system is able to control the aircraft using electric autopilot signals based on the control law created

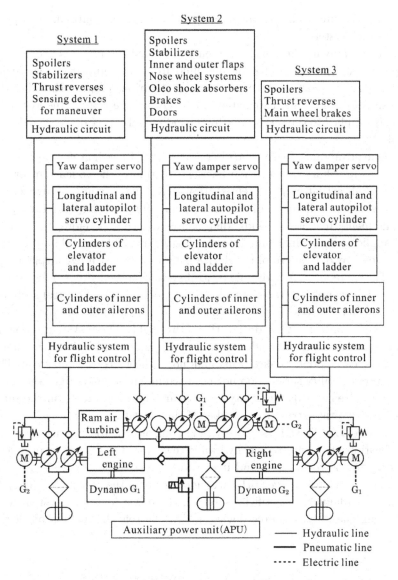

Fig. 1.13   Hydraulic control system of Boeing 767 passenger airplane.

from a computer. The fighter aircraft F-16 is the first airplane equipped with FBW system.

Aircraft hydraulic control systems are usually lumped source device type systems that supply hydraulic fluid from the lumped source device to the servo valves with actuators located at various places, using the long transmission lines. In a distributed source device type system, the hydraulic fluid is supplied to the servo valve by the individual source device fitted next to it.

One advantage of this type of system is that its weight can be reduced owing to the short transmission lines. Furthermore, the concurrent breakdown of all actuators is prevented. However, it has certain disadvantages, such as lower reliability of the individual source device and the space requirements to hold the individual source device. Therefore, the practical applications of distributed source devices in flight control systems are not widespread. The use of the device is limited to just a few airplanes.

There are aircraft hydraulic control systems intended not only for primary flight control systems but also to support operation control systems, called utility services, that are needed only during take-off and touch-down and maybe once or twice during the entire flight. These systems support vital operations such as undercarriage lowering and retraction, wing landing flaps, cargo doors brakes, wing fold, nose-wheel steering and so on. The actuators in utility services do not need redundant systems since the reliability requirements of the utility services are considered to be in the same class as any other structural member of the aircraft.

Figure 1.14 illustrates the attitude control of a rocket. Inclining the combustion chamber by the hydraulic actuator, the thrust is produced vertical to the rocket axis and it controls the attitude of the rocket. For instance, in the first stage of the attitude control of the rocket (NASDAJ-1) which is to launch an artificial satellite, there is a hydraulic system with an accumulator type source device, which controls the inclination of the

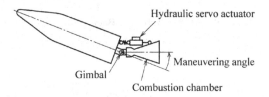

Fig. 1.14   Attitude control of rocket.

four combustion chambers of the outer veneer engines. In the attitude control of a rocket, the reduction in rigidity of the mechanism maintaining the combustion chamber often affects the control performance, which becomes a major difficulty.

The requirements for light weight, high reliability and durability of system components are predominant in the aviation fields. To make the hydraulic systems light weight, most component materials consist of aluminum alloys and alloy steels. In addition, stringent design procedures are demanded to ensure that the safety factor of components in relation to breaking strength of materials is about three. In aircraft hydraulic control systems it is usual to use redundant systems, aircraft fluid, aircraft seals and pressurized reservoirs in order to achieve a high reliability in the operation and durability of system elements.

Chapter 2

Fluid Mechanics for Hydraulic Control Systems

The analysis and synthesis of hydraulic components and systems are typically based on theories of fluid mechanics since a hydraulic control system is an energy transmission control system making use of hydraulic fluid. Although hydraulic fluid is a liquid, it holds air, either dissolved or in the form of suspended bubbles. The entrapped air affects the compressibility of the hydraulic fluid, but the fluid may still be treated as being incompressible except when considering the dynamic characteristics of hydraulic components and systems.

In this chapter, we are going to study the basic principles of fluid mechanics in hydraulic systems.

2.1 Fluid Properties and Hydraulic Fluids

2.1.1 *Fluid Properties and Definitions*

Let us consider a small element in a static fluid as shown in Fig. 2.1. It is maintained in a state of dynamic equilibrium by the forces normal to the surface of the element. Denoting the force normal to the surface of the element by ΔF and the surface area by ΔA, the pressure p is defined as follows.

$$p = \lim_{\Delta A \to 0} \frac{\Delta F}{\Delta A}. \tag{2.1}$$

In SI units (*Le Système International d'Unités*), the unit of a pressure is defined as Pascal [Pa], which is equivalent to N/m^2. The SI units of mass, length and time are defined as the kilogram [kg], meter [m] and second [s],

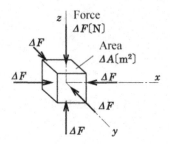

Fig. 2.1   Small element in static fluid.

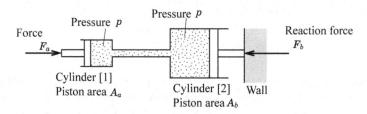

Fig. 2.2   Pressure transmission.

respectively, and the unit of force is defined as the Newton [N], which expresses a product of mass [kg] and the acceleration of gravity $(9.8\,\text{m/s}^2)$. As shown in Appendix, the prefixed symbols of SI units are represented in Table A.2 and the relationships between SI units and other units are given in Table A.3.

Pressure measured above absolute zero is referred to as the absolute pressure. The pressure of a standard atmosphere is 101.3 kPa in absolute pressure. Pressure measured relative to atmospheric pressure is called the gage pressure.

The well-known Pascal's law shows that pressure applied to a part of the fluid volume in a closed container is transmitted to all the fluid inside. Let us consider two cylinders connected by a transmission line as shown in Fig. 2.2. Applying a force F_a to the piston of the cylinder [1], the pressure $p = F_a/A_a$ is transmitted to the fluid in the cylinder [2]. Therefore, the force F_b acting on the wall becomes

$$F_b = pA_b = \frac{A_b}{A_a}F_a. \tag{2.2}$$

Fluid properties such as density, viscosity, compressibility and bulk modulus are defined as follows.

2.1.1.1 *Density*

Density ρ of a fluid is defined by the ratio of the fluid mass to its volume. The density of water, ρ is $1000\,\mathrm{kg/m^3}$ at the absolute pressure $101.3\,\mathrm{kPa}$ and temperature $4°\mathrm{C}$. The density ρ of a hydraulic mineral fluid is about 800–$900\,\mathrm{kg/m^3}$, and the empirical formula expressing the relationship between density $\rho[\mathrm{kg/m^3}]$ and pressure p [Pa] is given as follows:

$$\rho \approx \rho_0(1 + 6.40 \times 10^{-10}p), \tag{2.3}$$

where ρ_0 is density at the normal standard condition, and the pressure p denotes the gage pressure. All references to pressure in this book will refer to gage pressure unless specified otherwise.

For a temperature change of ΔT [°C] from the standard temperature, the empirical formula expressing the density ρ of hydraulic mineral fluid is given as follows:

$$\rho = \rho_0(1 - a\Delta T), \tag{2.4}$$

where coefficient a is $a = (6.3\text{–}8.7) \times 10^{-4°}\mathrm{C}^{-1}$.

2.1.1.2 *Viscosity and Kinematic Viscosity*

Viscosity is the property of a fluid expressing its resistance to shear or angular deformation. Consider some fluid flowing along a plate as shown in Fig. 2.3. The particles of the fluid in contact with the plate adhere to it, and the velocity of fluid particles increases as the distance y from the plate increases. Thus, a velocity gradient du/dy exists. The action may be regarded as if the fluid were made up of a series of thin sheets with

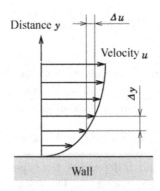

Fig. 2.3   Flow in boundary layer.

a shearing force acting between them. Then, if the shearing force f is on the surface between any two thin fluid sheets with area A, the shear stress $\tau = f/A$ can be expressed by

$$\tau = \mu \frac{du}{dy}. \tag{2.5}$$

Equation (2.5) is referred to as the Newton's law of viscosity, and the symbol μ is defined as viscosity, which is an inherent property of the fluid, independent of the velocity gradient du/dy.

The fluid to which Eq. (2.5) is applicable is referred to as Newtonian fluid, and most hydraulic fluids belong to this category, except some synthetic hydraulic fluids. Widely used viscosity units include the Poise [P] and centi-poise [cP], which are Centimeter-Gram-Second (CGS) system of units. A Poise is defined as the viscosity giving rise to the shearing stress $\tau = 1 \times 10^{-1} \mathrm{N/m^2}$ at $du = \Delta u = 1 \times 10^{-2} \mathrm{m/s}$, $dy = \Delta y = 1 \times 10^{-2} \mathrm{m}$ in Eq. (2.5).

Accordingly, we get:

$$1[\mathrm{P}] = \frac{\tau \Delta y}{\Delta u} = \frac{1 \times 10^{-1} \left[\frac{\mathrm{N}}{\mathrm{m^2}}\right] \times 1 \times 10^{-2}[\mathrm{m}]}{1 \times 10^{-2} \left[\frac{\mathrm{m}}{\mathrm{s}}\right]} = 0.1[\mathrm{Pa \cdot s}]. \tag{2.6}$$

Viscosity μ is also referred to as absolute viscosity to avoid confusion with kinematic viscosity $\nu = \mu/\rho$, which is the ratio of viscosity μ to density ρ. Widely used CGS units of kinematic viscosity include Stokes [St] and centistokes [cSt].

Kinematic viscosity of $1\,\mathrm{St}$ equals $1 \times 10^{-4} \mathrm{m^2/s}$ which corresponds to absolute viscosity 1 P in a fluid with density $1 \times 10^3 \mathrm{kg/m^3}$. Hence,

$$1[\mathrm{St}] = \frac{1[\mathrm{P}]}{1 \times 10^3 \left[\frac{\mathrm{kg}}{\mathrm{m^3}}\right]} = 1 \times 10^{-4} \left[\frac{\mathrm{N \cdot s}}{\mathrm{m^2}} \frac{\mathrm{m^3}}{\mathrm{kg}}\right] = 1 \times 10^{-4} \left[\frac{\mathrm{m^2}}{\mathrm{s}}\right]. \tag{2.7}$$

The viscosity of a hydraulic fluid changes with temperature, and this affects the performance of hydraulic systems. Walther's experimental formula yields the relationship between the kinematic viscosity and temperature:

$$\log\left[\log(\nu + 0.8)\right] = A - B \log T, \tag{2.8}$$

where the coefficients A and B depend on the type of hydraulic fluid, and T [K] is the thermodynamic temperature. The relationships between thermodynamic temperature T[K], Centigrade temperature t_C[°C] and Fahrenheit

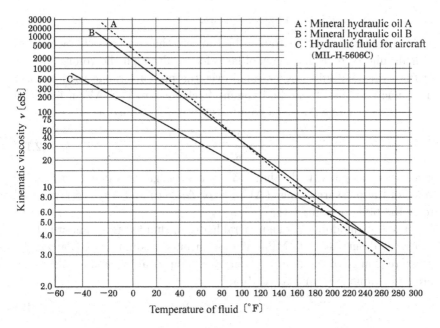

Fig. 2.4 ASTM standard viscosity–temperature chart for liquid petroleum products.

temperature $t_F[°F]$ scales are given by the formula:

$$T = t_C + 273.15 = \frac{t_F + 459.67}{1.8}. \tag{2.9}$$

Figure 2.4 is a part of the American Society for Testing and Materials (ASTM) standard viscosity–temperature chart for liquid petroleum products. As shown in Fig. 2.4, the linear relationships in the semi-logarithmic chart enable estimates to be made of fluid viscosity at any temperature in the range shown.

Fluid viscosity is affected very little by changes in pressure, although those changes may impact on the quality of hydraulic bearing lubrication. The empirical formula expressing the relationship between viscosity μ and the pressure change p is given as follows:

$$\frac{\mu}{\mu_0} = e^{\alpha p}, \tag{2.10}$$

where μ_0 is viscosity at the atmospheric pressure, and exponent α is given as $\alpha = (1.6 \sim 3.6) \times 10^{-8}\,\mathrm{Pa}^{-1}$ for mineral hydraulic fluids.

2.1.1.3 *Compressibility and Bulk Modulus*

Compressibility of the hydraulic fluid affects the dynamic characteristics of hydraulic systems that control large loads, although the volume change of the hydraulic fluid is very small over the pressure range used in most hydraulic systems. If the fluid volume V changes to $V + \Delta V$ in line with the pressure change Δp, then compressibility β is defined as

$$\beta = \frac{1}{\kappa} = -\frac{\frac{\Delta V}{V}}{\Delta p}, \tag{2.11}$$

where the change of fluid volume ΔV takes a negative value for a positive value of the pressure change ($\Delta V < 0$ for $\Delta p > 0$) and $\kappa = 1/\beta$ is referred to as the bulk modulus.

The bulk modulus κ_0 of a mineral hydraulic fluid at the standard atmosphere is about 1.6×10^3 MPa. The mean bulk modulus κ_m[MPa] for the pressure change of the fluid is given by an empirical formula applicable to pressures ranging from 0 to 70 MPa:

$$\kappa_m = \kappa_0 + 5.30p. \tag{2.12}$$

However, the bulk modulus of a real liquid differs from that of an ideal hydraulic fluid because the working fluid takes in air in the forms of entrapped small bubbles and dissolved air during operation of the hydraulic system. Moreover, elasticity of the container enclosing the hydraulic fluid leads to reduction of the bulk modulus of hydraulic fluids. The bulk modulus of a virtual hydraulic fluid is called the effective bulk modulus κ_e. The effective bulk modulus κ_e in typical hydraulic system falls in the range from 0.5×10^3 to 1×10^3 MPa.

(a) *Effective bulk modulus taking into account the elasticity of a fluid container*

Let us consider a hydraulic fluid placed in a container with volume V_c under pressure p. As the fluid pressure changes inside the container ($\Delta p > 0$), the container volume V_c becomes $V_c + \Delta V_c$, and the fluid volume V_0 in the container becomes $V_0 + \Delta V_0$ as the container is expanded and the fluid is compressed. Then, the net fluid volume change $-\Delta V_e > 0$ in the container is given by

$$-\Delta V_e = -\Delta V_0 + \Delta V_c \qquad (-\Delta V_0 > 0, \quad \Delta V_c > 0). \tag{2.13}$$

Regarding the expansion of the container, ΔV_c as the compressed volume change of the working fluid, the reciprocal of the effective bulk modulus taking into account the elasticity of a fluid container is given by Eq. (2.14):

$$\frac{1}{\kappa_e} = \frac{-\Delta V_e}{V_0 \Delta p} = \frac{-\Delta V_0}{V_0 \Delta p} + \frac{\Delta V_c}{V_c \Delta p} = \frac{1}{\kappa_0} + \frac{1}{\kappa_c}. \tag{2.14}$$

In the right-hand side of Eq. (2.14), the first term $\beta_0 = 1/\kappa_0$ expresses the compressibility of the hydraulic fluid and the second term $\beta_c = 1/\kappa_c$ is the expansion rate of the container.

For a cylindrical container with inner diameter d_i and outer diameter d_o, the expansion rate that is expressed by minus coefficient of compressibility β_c of the container is given as follows:

$$\beta_c = \frac{1}{\kappa_c} = \frac{2}{E}\left(\frac{d_o^2 + d_i^2}{d_o^2 - d_i^2} + \mu_p\right), \tag{2.15}$$

where μ_p is Poisson's ratio for the container material having a modulus of longitudinal elasticity E. Typical values of Poisson's ratio and the modulus of elasticity for steel are $\mu_p = 0.25$ and $E = 2.1 \times 10^{11}\,\mathrm{N/m^2}$, respectively.

Let us consider a container comprising two tubes made of different materials and with different sizes, as shown in Fig. 2.5. Denoting the fluid volumes in two tubes by V_A and V_B, respectively, the effective changes in fluid volume ΔV_A and ΔV_B resulting from a pressure change Δp are described as follows, using the effective bulk modulus κ_A for the fluid in tube A and κ_B for the fluid in tube B:

$$\left.\begin{aligned} -\Delta V_A &= \frac{V_A \Delta p}{\kappa_A} \\ -\Delta V_B &= \frac{V_B \Delta p}{\kappa_B} \end{aligned}\right\}. \tag{2.16}$$

Therefore, the effective bulk modulus of the container comprising tubes A and B is given by:

$$\frac{1}{\kappa_e} = \frac{-(\Delta V_A + \Delta V_B)}{(V_A + V_B)\Delta p} = \frac{V_A}{(V_A + V_B)}\frac{1}{\kappa_A} + \frac{V_B}{(V_A + V_B)}\frac{1}{\kappa_B}. \tag{2.17}$$

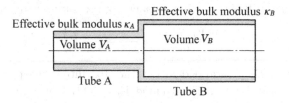

Fig. 2.5   Container comprising tubes A and B.

(b) *Effective bulk modulus of hydraulic fluid taking into account sus-*
    pended air

Hydraulic fluid contains from 7% to 10% air by volume, dissolved or
entrapped in the form of small bubbles. The entrapped air affects the bulk
modulus, but the influence of dissolved air on the bulk modulus may be
negligible.

In most cases, the equation of state for ideal gases is applicable to real
gases and it is given as

$$pV_a^n = \text{const.}, \tag{2.18}$$

where V_a is a specific gas volume (volume per unit density), p is a pressure
and n is an exponent number which takes the value 1.4 for isentropic (adiabatic) change of air, and 1.0 for isothermal processes. It is assumed that
in an isothermal process, all energy associated with the change of state is
transferred out of the gas by way of heat transfer. During adiabatic processes, it is assumed that all energy associated with the change of state is
held within the gas. Since the change of state of the gas generally involves
heat transfer to the environment, the air takes the index number n some-
where in the range from 1.0 to 1.4, referred to as the polytropic exponent
of air.

Let us now consider the effective bulk modulus κ_e of hydraulic fluid
with the entrapped air of the volume V_a. Differentiating Eq. (2.18) with
respect to the pressure p gives

$$dV_a = -\frac{V_a}{np}dp, \tag{2.19}$$

ΔV, Δp and κ in Eq. (2.11) being replaced by dV_f, dp and κ_f, the volume
change dV_f of the hydraulic fluid except the entrapped air is given as follows:

$$dV_f = -\frac{V_f}{\kappa_f}dp. \tag{2.20}$$

Recalling Eqs. (2.19) and (2.20), the effective bulk modulus κ_e of the
hydraulic fluid with entrapped air under the pressure p is described as

$$\frac{1}{\kappa_e} = -\frac{1}{(V_a + V_f)}\frac{d(V_a + V_f)}{dp} = \frac{1}{(V_a + V_f)}\left(\frac{V_a}{np} + \frac{V_f}{\kappa_f}\right). \tag{2.21}$$

The volume of the entrapped air at the absolute atmospheric pressure
p_0 being denoted by V_{ao}, the volume V_a of entrapped air at the absolute pressure p is described as $V_a = V_{ao}(p_0/p)^{1/n}$, in accordance with

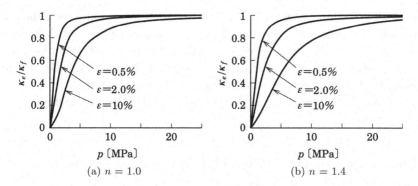

Fig. 2.6 Effective bulk modulus ratio κ_e/κ_f of hydraulic fluid with entrapped air.

Eq. (2.18). Therefore, the effective bulk modulus ratio (κ_e/κ_f) is derived from Eq. (2.21):

$$\frac{\kappa_e}{\kappa_f} = \frac{\left(\frac{p_0}{p}\right)^{\frac{1}{n}} \varepsilon + 1}{\left(\frac{p_0}{p}\right)^{\frac{1}{n}} \frac{\varepsilon \kappa_f}{np} + 1}, \tag{2.22}$$

where κ_f is the bulk modulus for a hydraulic fluid and ε is the volume ratio V_{ao}/V_f of entrapped air to hydraulic fluid. Figure 2.6 shows the relationship between the dimensionless effective bulk modulus κ_e/κ_f and pressure p for various values of the volume ratio ε when κ_f is a constant, with a value of 1.6×10^3 MPa. Figures 2.6(a) and 2.6(b) correspond to slow and rapid changes of pressure, i.e. isothermal processes with $n = 1.0$ and adiabatic processes with $n = 1.4$, respectively.

2.1.2 *Hydraulic Fluid*

Hydraulic fluid is not only a medium used to transmit power but also plays a major role in lubrication, resistance to rust, and dissipation of the heat generated by driving the system. Therefore, it is required that the other important properties of hydraulic fluid, such as viscosity, anti-wear, resistance to oxidation and rust, de-foaming and detergent dispersing, etc. should be precisely controlled. There are many kinds of hydraulic fluid suitable for hydraulic control systems, and the most appropriate hydraulic fluid can be selected to meet specific requirements. Hydraulic fluids used in most applications are petroleum-based oils, typically used in the temperature range from 20°C to 65°C. Anti-wear type mineral hydraulic fluids are commonly used now as hydraulic fluids for general industries, though rust and oxidation inhibited mineral hydraulic fluids have been widely used. Aircraft fluids need to be fire-resistant and be able to operate in a wide temperature range and at a pour point of $-60°C$ or less. Recently, hydraulic systems using water have been developed and used as low-pressure systems, for environmental reasons. There are several kinds of hydraulic fluids.

2.1.2.1 *Mineral Hydraulic Fluids*

Hydraulic fluids used in a majority of applications are petroleum-based oils.

(a) *Straight mineral oil*

Straight turbine oil (#90, #140) is used as the hydraulic fluid in low-pressure systems of 7 MPa or less.

(b) *R&O type hydraulic oil*

R&O type hydraulic oils are petroleum-based hydraulic fluids refined from selected crude oil and mixed with the additives to prevent rust, oxidation and foaming.

(c) *Anti-wear type hydraulic oil*

Anti-wear type hydraulic oils are R&O type hydraulic fluids with additives to prevent the wearing of the scrubbed surface, such as the inner surface of a cam ring in a vane pump.

(d) *Wide temperature range type hydraulic oil*

The wide temperature range type of hydraulic oils have the pour point at $-40°C$ or less and are usable at even $100°C$ or more. Hydraulic fluids used in aircraft (American military standard MIL-H-5606C, MIL-H83282A, etc.) belong to the group of wide temperature range type hydraulic oils.

2.1.2.2 *Synthetic Fluids*

Synthetic fluids are artificially synthesized from refined mineral oil to provide them with the required characteristics. Synthetic fluids cost several times more than general mineral hydraulic fluids.

(a) *Phosphate ester-based synthetic fluids*

Straight phosphate esters and oil synthetic blends have fire resistance and anti-wear qualities, and are excellent lubricants though their viscosity is lower than that of mineral hydraulic fluids. When this fluid is used, the seal material has to be made corrosion-resistant.

(b) *Fatty acid ester-based synthetic fluids*

Fatty acid ester-based synthetic fluids are fire-resistant, have an excellent anti-wear performance and are excellent lubricants. The fatty acid ester-based synthetic fluids have high viscosity compared with phosphate ester-based synthetic fluids, and general purpose hydraulic seals made of rubber materials can be used.

2.1.2.3 *Water Content Hydraulic Fluids*

Water content hydraulic fluids have been developed to reduce the costs of hydraulic fluids and to make them fire-resistant. The water content hydraulic fluids have certain drawbacks as their corrosion, rust resistance and lubricating properties are inferior to those displayed by mineral hydraulic fluids. Moreover, they tend to cause cavitation because of the low viscosity of the fluid.

(a) *Water glycol fluids*

Water glycol fluids consist of 35–50% water, 35–45% glycol and additives (polyester) to increase viscosity. The lubrication and viscosity/temperature properties of water glycol fluids are superior to those of other water containing fluids. The pressure limit for these fluids is 21 MPa or less.

(b) *Water-in-oil emulsions*

Water-in-oil emulsions are mixtures of oil and 1–2 μm water droplets surrounded by an oil film, and contains 35–50% water and 50–60% mineral oil. The pressure limit for these fluids is 21 MPa or less.

(c) *High water content fluids*

High water content fluids (HWCF) contain 90–95% water and 5–10% additives, such as viscosity improvers, anti-friction additives and inhibitors of rust and bacteria growth, as well as sludge inhibitors. The HWCF fluids have excellent fire resistance and heat transfer capabilities, and moreover their cost is about 20% of the petroleum-based hydraulic fluids. However, the pressure limit for these fluids is 7 MPa or less, and the temperature must be kept within the range from 5–49°C.

Table 2.1 Properties of typical hydraulic fluids.

	Wide range type		R&O type
	MIL-H-5606C	MIL-H-83282A	ISO VG 32
Density [kg/m^3]	830(at 15°C)	840(at 15°C)	867(at 15°C)
Kinematic viscosity [cSt]	500(at −40°C)	1900(at −40°C)	30.3(at 40°C)
	3.4(at 135°C)	2.3(at 135°C)	5.4(at 100°C)
Working temperature	From −54 to	From −54 to	Pour point −35°C
range	135°C	135°C	Flash point 222°C

2.1.2.4 *City Water, River Water and Seawater*

Water is an environment-friendly, non-flammable and easily available fluid. The hydraulic system using water as hydraulic fluid is called a water hydraulic system and has been developed in the last decade. Water hydraulic systems have proved to be the best technology alternative in certain sectors of industry, such as the food processing, pharmacology, paper and steel production.

Water has lower viscosity and is an inferior lubricant compared to other hydraulic fluids. Therefore, water hydraulic components should have small, precisely controlled slight clearances between the moving parts to prevent leakages due to the low viscosity, without inhibiting lubrication. Other drawbacks include the easy growth of rust, corrosions and sludge. Therefore, high-pressure water hydraulic systems are not only expensive but unreliable. However, low-pressure water hydraulic systems are an attractive technology in sectors requiring a clean environment: such as food processing and nursing, where they have been put to good practical use. The main challenges for the water hydraulic technology are the improvement of reliability and controllability, while making them cost-effective.

Table 2.1 compiles the properties of typical hydraulic fluids and the aircraft fluids in the American military standard MIL-H-5606C and MIL-H-83282A.

2.2 Fluid Flow Concepts and Basic Equations

2.2.1 *Equations of Fluid Flow*

Considering the movement of fluid particles in a flow field, streamlines are defined by a series of continuous curves drawn so as to be tangential to velocity vectors of the particles at a given time instant. Flows are usually illustrated graphically by the streamlines. Since there are no fluid particles passing across the streamlines, the flow in an imaginary tube surrounded by the streamlines may be treated as a flow in a tube with no wall thickness. The imaginary tube is called a stream tube.

Denoting the flow rate passing through the cross-section of a small stream tube by dQ and fluid density by ρ, the mass flow rate is described by ρdQ, which is the mass per unit time passing through it. If the states of flow such as the velocity v, density ρ, pressure p and temperature T do not change with time t at any point in the flow field, the flow is referred to as steady flow.

Hence, the steady state flow yields $\partial v/\partial t = 0$, $\partial \rho/\partial t = 0$, $\partial p/\partial t = 0$ and $\partial T/\partial t = 0$, where the space ($xyz$ coordinates in the flow field) is held constant. The flow is referred to as unsteady flow when the state of the fluid flow changes with time at any point.

The velocity components u_x, u_y and u_z parallel to the coordinate axes at the point (x, y, z) at the time t are described by the functions of independent variables x, y, z, t:

$$\left. \begin{aligned} u_x &= f_x(x, y, z, t) \\ u_y &= f_y(x, y, z, t) \\ u_z &= f_z(x, y, z, t) \end{aligned} \right\}. \tag{2.23}$$

Supposing that an infinitesimal fluid element transfers from the coordinate (x, y, z) to $(x + \Delta x, y + \Delta y, z + \Delta z)$ through a small time change Δt, then the velocity change Δu_z of the fluid element is given as

$$\Delta u_x = u_x(x + \Delta x, y + \Delta y, z + \Delta z, t + \Delta t) - u_x(x, y, z, t). \tag{2.24}$$

Expanding the first right-hand side term of Eq. (2.24) into a Taylor's series, and neglecting the terms with a higher exponent than two yields the following equation:

$$\Delta u_x = \frac{\partial u_x}{\partial x}\Delta x + \frac{\partial u_x}{dy}\Delta y + \frac{\partial u_x}{\partial z}\Delta z + \frac{\partial u_x}{\partial t}\Delta t. \tag{2.25}$$

The equations expressing the velocity changes Δu_y and Δu_z are obtained in a similar manner. Therefore, the acceleration components ($a_x = \Delta u_x/\Delta t$,

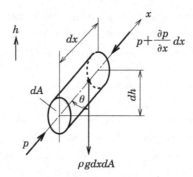

Fig. 2.7 Force acting on infinitesimal fluid element.

$a_y = \Delta u_y/\Delta t$, $a_z = \Delta u_z/\Delta t$ at $\Delta t \to 0$) are given as follows:

$$\left.\begin{aligned}
a_x &= \frac{du_x}{dt} = u_x\frac{\partial u_x}{\partial x} + u_y\frac{\partial u_x}{\partial y} + u_z\frac{\partial u_x}{\partial z} + \frac{\partial u_x}{\partial t} \\
a_y &= \frac{du_y}{dt} = u_x\frac{\partial u_y}{\partial x} + u_y\frac{\partial u_y}{\partial y} + u_z\frac{\partial u_y}{\partial z} + \frac{\partial u_y}{\partial t} \\
a_z &= \frac{du_z}{dt} = u_x\frac{\partial u_z}{\partial x} + u_y\frac{\partial u_z}{\partial z} + u_z\frac{\partial u_z}{\partial z} + \frac{\partial u_z}{\partial t}
\end{aligned}\right\}. \tag{2.26}$$

Now, for simplicity, consider a frictionless and one-dimensional flow defined by an x-coordinate position and time. Figure 2.7 illustrates an infinitesimal fluid column with a cross-section dA and length dx in the flow field. The external forces acting on the fluid column involve surface forces due to the pressure p and a body force due to gravity. The surface force on the upstream face with area dA is described by pdA for the x-direction, while the surface force for the x-direction on the downstream face with the area dA is described by $-[p + (\partial p/\partial x)dx]dA$.

The body force component for the x-direction is described by $-\rho dx dA \cdot g\cos\theta$, where g and θ are acceleration of gravity and an angle between the h-direction and x-direction. Applying Newton's law of motion to the infinitesimal fluid column gives

$$pdA - \left(p + \frac{\partial p}{\partial x}dx\right)dA - \rho g(\cos\theta)dxdA = \rho dxdA\frac{du_x}{dt}. \tag{2.27}$$

Equation (2.27) divided by $dxdA$ yields the equation of motion of the fluid element per unit volume:

$$-\frac{\partial p}{\partial x} - \rho g\cos\theta = \rho\frac{du_x}{dt}. \tag{2.28}$$

For one-dimensional flow, Eq. (2.26) is expressed as follows:

$$\frac{du_x}{dt} = u_x \frac{\partial u_x}{\partial x} + \frac{\partial u_x}{\partial t}. \tag{2.29}$$

Equation (2.28) may be rearranged using Eq. (2.29):

$$\rho \left(\frac{\partial u_x}{\partial t} + u_x \frac{\partial u_x}{\partial x} \right) = -\rho g \cos \theta - \frac{\partial p}{\partial x}. \tag{2.30}$$

The left-hand side of the Eq. (2.30) expresses the inertia force acting on the fluid per unit volume. The first term on the right-hand side of the Eq. (2.30) expresses the body force acting on the fluid per unit volume, and the second term is the surface force acting on it.

In the case of a steady and incompressible flow, the term $\partial u_x / \partial t$ in Eq. (2.30) becomes zero and ρ is constant. Integrating both side of Eq. (2.30) with respect to the distance x gives

$$\rho \int u_x du_x = -\rho g \cos \theta \int dx - \int dp, \tag{2.31}$$

$$\therefore \quad \frac{1}{2}\rho u_x^2 + \rho g h + p = \text{const.}, \tag{2.32}$$

where $h (= \rho x \cos \theta)$ is a vertical distance.

Equation (2.32) is Bernoulli's equation of incompressible flow. The first term $(1/2)\rho u_x^2$ and the second term $\rho g h$ in Eq. (2.32) stand for the kinetic energy of the fluid per unit volume and potential energy of the fluid per unit volume, respectively. The last term expresses the pressure energy of the fluid per unit volume. In accordance with the Bernoulli's equation the whole energy of the fluid per unit volume consists of kinetic energy, potential energy and pressure energy, and the whole energy is constant in an incompressible frictionless flow.

Let us consider the three-dimensional motion of an infinitesimal frictionless fluid with a rectangular parallelepiped volume as shown in Fig. 2.8. The equations of motion of the fluid per unit volume corresponding to Eq. (2.30) are as follows:

$$\left. \begin{array}{l} \rho \left(u_x \dfrac{\partial u_x}{\partial x} + u_y \dfrac{\partial u_x}{\partial y} + u_z \dfrac{\partial u_x}{\partial z} + \dfrac{\partial u_x}{\partial t} \right) = \rho X - \dfrac{\partial p}{\partial x} \\[2ex] \rho \left(u_x \dfrac{\partial u_y}{\partial x} + u_y \dfrac{\partial u_y}{\partial y} + u_z \dfrac{\partial u_y}{\partial z} + \dfrac{\partial u_y}{\partial t} \right) = \rho Y - \dfrac{\partial p}{\partial y} \\[2ex] \rho \left(u_x \dfrac{\partial u_z}{\partial x} + u_y \dfrac{\partial u_z}{\partial y} + u_z \dfrac{\partial u_z}{\partial z} + \dfrac{\partial u_z}{\partial t} \right) = \rho Z - \dfrac{\partial p}{\partial z} \end{array} \right\}, \tag{2.33}$$

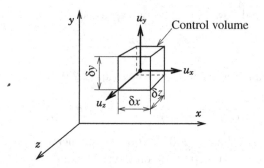

Fig. 2.8　Infinitesimal fluid element in three-dimensional flows.

where, X, Y and Z denote the body forces acting on the fluid per unit volume for x, y and z-coordinate directions, respectively. Equation (2.33) is Euler's equation of motion. Equations of motion for real fluids should take into account the forces generated due to fluid viscosity. As the procedure of deriving these equations, referred to as Navier–Stokes equations, is beyond the scope of this book, the equation for incompressible flow will be just quoted here:

$$
\left.
\begin{aligned}
& \rho\left(u_x\frac{\partial u_x}{\partial x} + u_y\frac{\partial u_x}{\partial y} + u_z\frac{\partial u_x}{\partial z} + \frac{\partial u_x}{\partial t}\right) \\
& \quad = \rho X - \frac{\partial p}{\partial x} + \mu\left(\nabla^2 u_x + \frac{1}{3}\frac{\partial}{\partial x}\mathrm{div}u\right) \\
& \rho\left(u_x\frac{\partial u_y}{\partial x} + u_y\frac{\partial u_y}{\partial y} + u_z\frac{\partial u_y}{\partial z} + \frac{\partial u_y}{\partial t}\right) \\
& \quad = \rho Y - \frac{\partial p}{\partial y} + \mu\left(\nabla^2 u_y + \frac{1}{3}\frac{\partial}{\partial y}\mathrm{div}u\right) \\
& \rho\left(u_x\frac{\partial u_z}{\partial x} + u_y\frac{\partial u_z}{\partial y} + u_z\frac{\partial u_z}{\partial z} + \frac{\partial u_z}{\partial t}\right) \\
& \quad = \rho Z - \frac{\partial p}{\partial z} + \mu\left(\nabla^2 u_z + \frac{1}{3}\frac{\partial}{\partial z}\mathrm{div}u\right)
\end{aligned}
\right\}, \tag{2.34}
$$

where, μ is viscosity of the fluid. The last term of right-hand side in Eq. (2.34) is the force acting on the fluid per unit volume depending on viscosity:

$$
\nabla^2 = \frac{\partial^2}{\partial x^2} + \frac{\partial^2}{\partial y^2} + \frac{\partial^2}{\partial z^2}, \tag{2.35}
$$

$$
\mathrm{div}u = \frac{\partial u_x}{\partial x} + \frac{\partial u_y}{\partial y} + \frac{\partial u_z}{\partial z}. \tag{2.36}
$$

2.2.2 *Equation of Continuity*

Fundamental principles in fluid mechanics such as continuity and momentum are derived from concepts of a system and a control volume. A system is an identified quantity of fluid in the flow field. The boundary of the system makes a closed surface, and its shape, position and thermal condition varies with a time, though it preserves the same mass. A control volume is a definite volume designated in the flow field. The size, shape and position of a control volume are fixed in the flow field, though the amount and identity of the fluid in the control volume may vary with a time. The boundary of a control volume is called a control surface.

Figure 2.9 shows a system whose boundary coincides exactly with a control volume at the time instant $t = t_0$ in the flow field. The fluid mass $m_v(t)$ in the control volume varies with time t because the fluid continually flows into and out of the control volume. Further, regarding the lump of fluid in the control volume at the time t_0 as a system, the fluid mass $m_v(t_0)$ in the control volume is equal to the system mass m.

The system has moved somewhat during the time change Δt and then it comprises the broken line volume shown in Fig. 2.9. The fluid mass $m_v(t_0 + \Delta t)$ in the control volume at the time $t_0 + \Delta t$ is described as follows:

$$m_v(t_0 + \Delta t) = m + \Delta m_{\text{in}} - \Delta m_{\text{out}}, \qquad (2.37)$$

where Δm_{in} stands for the fluid mass freshly flowing into the control volume during Δt, and Δm_{out} is the mass of the part of the system moving outside

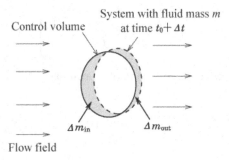

Fluid mass $m_v(t_0)$ in control volume at time t_0 = System with fluid mass m

$m_v(t_0 + \Delta t)$: Fluid mass in control volume at time $t_0 + \Delta t$

Fig. 2.9   Control volume and a system whose volume at time instant t_0 is identical to control volume.

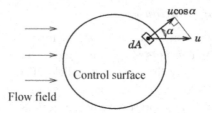

Fig. 2.10   Control surface and the elemental area dA.

the control volume during the time Δt. Therefore, Eq. (2.37) is arranged as

$$\frac{m_v\,(t_0 + \Delta t) - m}{\Delta t} + \frac{\Delta m_{\text{out}} - \Delta m_{\text{in}}}{\Delta t} = 0, \tag{2.38}$$

where $m = m_v(t_0)$. In Eq. (2.38), the first left-hand side term expresses the rate of the change of the mass within the control volume in time, and the second term expresses the net efflux rate of mass across the control volume boundary.

As shown in Fig. 2.10, the mass flow rate ρdQ through the area of dA is described by $\rho u \cos \alpha \cdot dA$, where $u \cos \alpha$ denotes the fluid velocity normal to the infinitesimal area dA on the control surface. Regarding the flow rate out of the infinitesimal area dA on the control volume as a plus vector, integrating the mass flow rate ρdQ on the whole control surface gives the rate of net efflux mass across the control volume boundary. Therefore, Eq. (2.38) can be arranged as follows:

$$\left.\begin{aligned}
\frac{dm_v(t)}{dt} + \int_A \rho u \cos \alpha\, dA = 0 \\[2mm]
\text{or} \quad \frac{dm_v(t)}{dt} + \int_A \rho dQ = 0
\end{aligned}\right\}. \tag{2.39}$$

Equation (2.39) is called a condition of continuity, and for the steady flow, it is given as

$$\int_A \rho u \cos \alpha\, dA = 0. \tag{2.40}$$

Figure 2.11 illustrates the fluid flow in a minute stream tube with the inlet area dA_1 and outlet area dA_2. Applying Eq. (2.40) to the steady flow in a minute stream tube gives

$$\rho_1 u_1 dA_1 = \rho_2 u_2 dA_2, \tag{2.41}$$

Fig. 2.11   Flow in a stream tube.

where u_1 and u_2 denote the fluid velocities at the inlet and outlet, respectively, and $dm_v(t)/dt$ is zero provided that the flow is in a steady state. Similarly, the condition of continuity for a steady flow with a mean velocity u_1 at the inlet area A_1 and mean velocity u_2 at the outlet area A_2 in a stream tube is described as follows:

$$A_1 u_1 = A_2 u_2. \tag{2.42}$$

This is called continuity equation. Regarding the minute rectangular parallelepiped volume $\delta x \delta y \delta z$ shown in Fig. 2.8 as a control volume in the three-dimensional flow, the first term in Eq. (2.39) is expressed as

$$\frac{dm_v(t)}{dt} = \frac{\partial(\rho \delta x \delta y \delta z)}{\partial t}. \tag{2.43}$$

While, the second term in Eq. (2.39) is given as

$$\int_A \rho dQ = \int_A \rho u \cos \alpha dA. \tag{2.44}$$

That is,

The second term

$$= [(\text{influx mass flow rate}) - (\text{outward mass flow rate})]_{\text{control surface}}.$$

The distances from the center point of the control volume $\delta x \delta y \delta z$ to the yz-planes, zx-planes and xy-planes are expressed by $\pm\delta x/2$, $\pm\delta y/2$ and $\pm\delta z/2$, respectively. Therefore, the net efflux rates of mass for each plane of the control volume, expressed by the second term in Eq. (2.39), are described as follows.

The net efflux rate of mass for yz-planes of the control volume:

$$\left[\int_A \rho u \cos \alpha dA \right]_{yz} = \left[\rho u_x + \frac{\partial(\rho u_x)}{\partial x}\frac{\delta x}{2} \right] \delta y \delta z - \left[\rho u_x - \frac{\partial(\rho u_x)}{\partial x}\frac{\delta x}{2} \right] \delta y \delta z$$

$$= \frac{\partial(\rho u_x)}{\partial x} \delta x \delta y \delta z. \tag{2.45}$$

For xz-planes,

$$\left[\int_A \rho u \cos \alpha dA \right]_{xz} = \frac{\partial(\rho u_y)}{\partial y} \delta x \delta y \delta z. \tag{2.46}$$

For xy-planes,

$$\left[\int_A \rho u \cos \alpha dA \right]_{xy} = \frac{\partial(\rho u_z)}{\partial z} \delta x \delta y \delta z, \tag{2.47}$$

where u_x, u_y and u_z denote the fluid velocity components parallel to the coordinate axes at the center point of the control volume. Regarding the minute rectangular parallelepiped fluid element in Fig. 2.8 as a control volume, Eq. (2.44) can be expressed by the summation of the net mass efflux flow rates for each plane. Substituting the net flux rate of mass passing through the control surface and the time rate of the mass change within the control volume shown by Eq. (2.43) into Eq. (2.39) gives,

$$\frac{\partial \rho}{\partial t} + \frac{\partial(\rho u_x)}{\partial x} + \frac{\partial(\rho u_y)}{\partial y} + \frac{\partial(\rho u_z)}{\partial z} = 0. \tag{2.48}$$

Equation (2.48) is called the equation of continuity.

2.2.3 *Momentum Theory*

Let us treat a fluid element occupying a control volume at time t_0 as a system in a flow field. The system with mass m has moved somewhat during the time change Δt and now comprises the volume enclosed by the broken line as shown in Fig. 2.12(a). Applying Newton's equation of motion for the system gives

$$F = \frac{d(mu)}{dt} = \frac{dM(t)}{dt} = \lim_{\Delta t \to 0} \frac{M(t_0 + \Delta t) - M(t_0)}{\Delta t}, \tag{2.49}$$

where F denotes the external force acting on the system and $M(t)$ stands for the momentum of the system. The momentum $M_v(t)$ of the fluid in the control volume also varies with the time t because the fluid continually flows into the control volume and out of it. By virtue of the definition of the system and the control volume in Fig. 2.12(a), the momentum $M_v(t)$ equals $M(t)$ at time $t = t_0$:

$$M_v(t_0) = M(t_0). \tag{2.50}$$

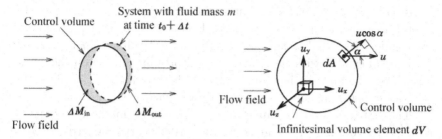

(a) Control volume and system   (b) Flow passing through control volume

Fig. 2.12   Momentum of fluid in control volume and momentum of a system whose volume at $t = t_0$ is identical to control volume.

At the time instant $t = t_0 + \Delta t$, the momentum $M_v(t_0 + \Delta t)$ of fluid in the control volume may be described as

$$M_v(t_0 + \Delta t) = M(t_0 + \Delta t) + \Delta M_{in} - \Delta M_{out}, \qquad (2.51)$$

where ΔM_{out} expresses the momentum of the part of the system moving out of the control volume during the infinitesimal time change Δt and ΔM_{in} stands for the momentum of the mass flowing into the control volume during the time change Δt as shown in Fig. 2.12(a). Equation (2.49) may be rearranged with recourse to Eqs. (2.50) and (2.51):

$$F = \lim_{\Delta t \to 0} \left[\frac{M_v(t_0 + \Delta t) - M_v(t_0)}{\Delta t} + \frac{\Delta M_{out} - \Delta M_{in}}{\Delta t} \right]. \qquad (2.52)$$

Equation (2.52) shows that the force acting on the system is given by adding the time-differentiated momentum of fluid within the control volume to the momentum of the net efflux fluid mass across the control surface per unit time.

Let us consider a limited size control volume comprising infinitesimal volume elements dV with the individual momentum as shown in Fig. 2.12(b) and the force F acting on the limited size system is described as:

$$F = \frac{dM_v}{dt} + \int_A \rho u \, dQ = \frac{d}{dt} \iiint_{c.v.} u\rho \, dV + \iint_{c.s.} u\rho(u \cos \alpha) dA, \qquad (2.53)$$

where $\rho dQ = (\rho u \cos \alpha)dA$ is a mass flow rate passing through the elemental area dA on the control surface. This is a vector that has negative magnitude when fluid is flowing into the control volume. In Eq. (2.53), the integral symbol with the suffix C.V. denotes integration on the whole control volume and the integral symbol with the suffix C.S. denotes the integration performed on the whole control surface.

Therefore, the force components F_x, F_y and F_z in the directions of x-, y- and z-coordinates are given by the formula further referred to as a momentum theory:

$$\left.\begin{array}{l} F_x = \dfrac{d}{dt} \iiint\limits_{\text{c.v.}} u_x \rho dV + \iint\limits_{\text{c.s.}} u_x \rho (u \cos \alpha) dA \\[2em] F_y = \dfrac{d}{dt} \iiint\limits_{\text{c.v.}} u_y \rho dV + \iint\limits_{\text{c.s.}} u_y \rho (u \cos \alpha) dA \\[2em] F_z = \dfrac{d}{dt} \iiint\limits_{\text{c.v.}} u_z \rho dV + \iint\limits_{\text{c.s.}} u_z \rho (u \cos \alpha) dA \end{array}\right\}. \qquad (2.54)$$

Equations (2.39) and (2.54) provide an approach for solving many problems in fluid mechanics. Since the momentum theory is a relationship between forces and momentums, we may choose an appropriate control volume involving those forces and momentums that will contribute to the solution of the problem. For instance, considering the case of a frictionless fluid flowing at a steady flow rate Q in a straight pipe with the inlet cross-section a_1 and outlet cross-section a_2, we may choose the inner volume of the pipe as the control volume to solve the force acting on the straight pipe. Then, the force f acting on the straight pipe surface is regarded as the reaction of the force acting on the fluid in the straight pipe. Since it may be assumed that there is no pressure difference between the inlet and outlet of the pipe by virtue of the frictionless flow assumption, applying Eq. (2.54) to the fluid in the control volume gives

$$f = -F = \rho Q u_1 - \rho Q u_2 = \frac{\rho Q^2}{a_1} - \frac{\rho Q^2}{a_2}, \qquad (2.55)$$

where $u_1 = Q/a_1$ and $u_2 = Q/a_2$ are mean velocity normal to the cross-section at the inlet and at the outlet, respectively, the force f is a vector with the positive magnitude in the direction of the flow, ρ is the fluid density and the time differentiation of the momentum within the control volume is zero by virtue of the assumption of steady flow.

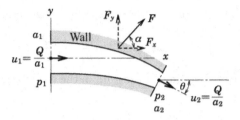

Fig. 2.13 Force acting on tube wall.

An application of momentum theory and the equation of continuity to a practical problem is illustrated in the following example.

[Example 2.1]

Figure 2.13 shows a reducing pipe with steady flow. The outlet flow direction is inclined at the angle θ for the inlet flow direction, where a is the cross-section areas of the pipe, p is the mean value of the pressures at the cross-section, and the mean velocities u at the cross-section are denoted by the suffixes 1 and 2 for the locations at the inlet and the outlet, respectively. Derive the equation of the force F acting on the reducing pipe.

[Solution 2.1]

The force F acting on the inner surface of the reducing pipe may be resolved into the x-direction component F_x and y-direction component F_y, hence it is described as follows:

$$F = \sqrt{F_x^2 + F_y^2}. \tag{1}$$

The x-direction force component acting on the fluid within the reducing pipe is described as

$$f_x = p_1 a_1 - p_2 a_2 \cos \theta - F_x. \tag{2}$$

Now, let us consider the volume in the reducing pipe as a control volume. Then, applying Eq. (2.54) to the fluid in the control volume yields the x-direction force component f_x acting on the fluid in the reducing pipe as follows:

$$f_x = \rho Q u_2 \cos \theta - \rho Q u_1 = \rho \frac{Q^2}{a_2} \cos \theta - \rho \frac{Q^2}{a_1}. \tag{3}$$

Cancelling f_x in Eqs. (2) and (3) gives the equation governing the force F_x:

$$F_x = p_1 a_1 - p_2 a_2 \cos\theta + \rho\frac{Q^2}{a_1} - \rho\frac{Q^2}{a_2}\cos\theta. \qquad (4)$$

The force component f_y acting on the fluid in the reducing pipe is described as:

$$f_y = p_2 a_2 \sin\theta - F_y. \qquad (5)$$

Applying Eq. (2.54) to the fluid in the control volume gives the y-direction force component f_y as follows:

$$f_y = -\rho Q u_2 \sin\theta. \qquad (6)$$

Substituting Eq. (6) into Eq. (5) gives

$$F_y = \left(\frac{\rho Q^2}{a_2} + p_2 a_2\right)\sin\theta. \qquad (7)$$

We obtain the following equations by using Eqs. (1), (4) and (7):

$$F = \sqrt{\left[p_1 a_1 - p_2 a_2 \cos\theta + \rho Q^2\left(\frac{1}{a_1} - \frac{\cos\theta}{a_2}\right)\right]^2 + \left(\frac{\rho Q^2}{a_2} + p_2 a_2\right)^2 \sin^2\theta},$$

$$(8)$$

$$\alpha = \tan^{-1}\frac{F_y}{F_x}. \qquad (9)$$

2.3 Inner Flow in Hydraulic Systems

2.3.1 *Flow Passing Through a Throttle*

When fluid is supplied to a hydraulic system, a certain amount of the energy contained in the fluid is used to overcome fluid friction. The major energy losses occur at throttles such as the orifices and chokes in the system. An orifice is a throttle so thin that friction losses in the orifice are negligible. On the other hand, a choke is a throttle with sufficient thickness to create an energy loss due to friction. The energy loss through the choke can be estimated using Eq. (2.75) shown in Section 2.3.3 because flow in a choke is generally laminar.

Figure 2.14 shows how flow is configured in a pipeline with a circular orifice which has a knife-edge configuration. Far upstream of the orifice, the fluid flows parallel to the center axis of the pipeline, but the flow contracts rapidly as it approaches the orifice. The cross-sectional area of the flow

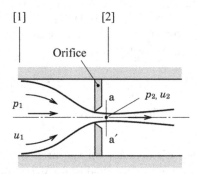

Fig. 2.14   Flow passing through an orifice.

path reaches a minimum at Section a–a' located just downstream of the orifice opening, and then increases gradually back to the full diameter of the pipeline.

The conservation of mass shows that the mean velocity increases with the decrease of the flow cross-sectional area, as given by the continuity equation (2.42). The conservation of energy is governed by the Bernoulli's equation (2.32) which implies that the pressure energy in a contracting flow is transformed into kinetic energy. Energy loss due to fluid friction is negligibly small in contracting flow because the fluid inertia force is far greater than the shear force. Applying Eq. (2.32) to a flow through the orifice gives

$$\frac{\rho u_1^2}{2} + p_1 = \frac{\rho u_2^2}{2} + p_2, \tag{2.56}$$

where u and p denote the mean fluid velocities and the mean pressures at the given cross-section, and ρ denotes the fluid density. The suffixes 1 and 2 signify the inlet [1] and the outlet [2] of the flow path, respectively. Substituting the continuity equation, i.e. Eq. (2.42) into Eq. (2.56) gives:

$$u_2 = \sqrt{\frac{2(p_1 - p_2)}{\rho\left[1 - \left(\frac{A_2}{A_1}\right)^2\right]}}. \tag{2.57}$$

Equation (2.57) may be approximated to Eq. (2.58) when the area ratio A_2/A_1 is extremely small:

$$u_2 = \sqrt{\frac{2(p_1 - p_2)}{\rho}}. \tag{2.58}$$

The flow cross-sectional area A_2 at a–a' is slightly smaller than the orifice opening area A_0. However, for convenience, we may describe a flow rate q through an orifice using the orifice opening area A_0 together with a flow coefficient C_d:

$$q = C_d A_0 u_2 = C_d A_0 \sqrt{\frac{2(p_1 - p_2)}{\rho}}. \qquad (2.59)$$

The flow coefficient C_d depends on the area ratio A_1/A_0, the orifice shape and Reynolds number as described later. Values of the flow coefficient C_d have been experimentally found to lie in the range 0.6 to 0.8 except for flows at very small Reynolds numbers.

The boundary in the flow through the cross-section a–a' grows gradually by absorbing fluid particles surrounding the flow path. When they are mixed, the kinetic energies of the fluids are transformed into the heat energy due to the fluid friction almost without recovering the pressure. By considering the unrecoverable pressure Δp that corresponds to the heat energy $\rho c_p \Delta T$ per unit volume of fluid, the Bernoulli's equation (2.32) may be rearranged as follows:

$$\frac{\rho u_1^2}{2} + p_1 + \rho g h_1 = \frac{\rho u_3^2}{2} + p_3 + \rho g h_3 + \rho c_p \Delta T, \qquad (2.60)$$

where c_p is the specific heat of the fluid (which means the energy required to raise the temperature of a unit mass of the fluid by $1°C$), and ΔT is the temperature difference between the locations indicated with suffix 1 and suffix 3, where section [3] is located well downstream of the orifice.

2.3.2 *Flow Rates Into and Out of a Cylinder*

Consider a vessel with volume V as shown in Fig. 2.15, in which the mass flow rate of fluid ρQ_1 flows into the vessel, and the mass flow rate ρQ_2 flows out of the vessel. Applying Eq. (2.39) to the flow through the vessel gives the relationship between the flow rates Q_1, Q_2 and the fluid mass ρV in

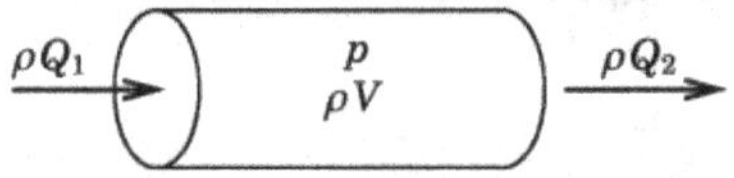

Fig. 2.15   Flow rates into and out of a vessel with volume V.

the vessel as follows:

$$\rho Q_1 - \rho Q_2 = \frac{d(\rho V)}{dt} \tag{2.61}$$

Therefore

$$Q_1 - Q_2 = \frac{dV}{dt} + \frac{V}{\rho}\frac{d\rho}{dt}. \tag{2.62}$$

Considering the fluid in the vessel as a system, the fluid mass ρV is as follows:

$$\rho V = \text{const.} \tag{2.63}$$

Differentiating the Eq. (2.63) with respect to the volume V gives

$$\frac{d\rho}{dV} = -\frac{\rho}{V}, \tag{2.64}$$

where dV is described as $-(1/\kappa)Vdp$ by virtue of the definition of the bulk modulus shown in Eq. (2.11). Using Eq. (2.64) and $dV = -(1/\kappa)Vdp$, the second term $(V/\rho)(d\rho/dt)$ in the right-hand side of Eq. (2.62) may be arranged as follows:

$$\frac{V}{\rho}\frac{d\rho}{dt} = \frac{V}{\kappa}\frac{dp}{dt}. \tag{2.65}$$

Substituting Eq. (2.65) into Eq. (2.62) gives

$$Q_1 - Q_2 = \frac{dV}{dt} + \frac{V}{\kappa}\frac{dp}{dt}. \tag{2.66}$$

Equation (2.66) is the flow continuity in which both the compressibility of the fluid and the change in volume of the vessel are accounted for. The first term on the right-hand side of Eq. (2.66) denotes the part of the net efflux volume flow rate which is associated with the change in volume of the vessel, and the second term denotes the other part, associated with fluid compressibility.

Let us now consider an application of Eq. (2.66) to the hydraulic cylinder with cross-sectional area A as shown in Fig. 2.16.

The flow rate of fluid into the right cylinder chamber with volume V_1 through port [1] is q_1 and the flow rate out of the left cylinder chamber

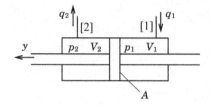

Fig. 2.16   Flow rates into and out of cylinder chambers.

with volume V_2 through port [2] is q_2. Therefore, applying Eq. (2.66) to the fluid in the right chamber gives

$$q_1 - 0 = \frac{dV_1}{dt} + \frac{V_1}{\kappa}\frac{dp_1}{dt}. \qquad (2.67)$$

Applying Eq. (2.66) to fluid in the left chamber gives

$$0 - q_2 = \frac{dV_2}{dt} + \frac{V_2}{\kappa}\frac{dp_2}{dt}. \qquad (2.68)$$

Let us consider the case when there is a small piston displacement y at the center piston position, where each cylinder volume has a volume V_0. Then the volume of fluid in the right side chamber volume V_1 changes from V_0 to $V_1 = V_0 + Ay$ and the volume of fluid in the left side chamber V_2 changes from V_0 to $V_2 = V_0 - Ay$. Then, Eqs. (2.67) and (2.68) are arranged as follows:

$$q_1 = A\frac{dy}{dt} + \frac{(V_0 + Ay)}{\kappa}\frac{dp_1}{dt}, \qquad (2.69)$$

$$q_2 = A\frac{dy}{dt} - \frac{(V_0 - Ay)}{\kappa}\frac{dp_2}{dt}. \qquad (2.70)$$

Equation (2.69) and (2.70) express the flow rates into and out of the cylinder, respectively, and they may be used in an analysis of the dynamic behavior of a hydraulic cylinder system.

2.3.3 *Flow in a Pipeline*

Visualizing the water flow by injecting ink into a glass pipe from a fine tube as shown in Fig. 2.17, we can observe that the ink flows in the form of a line without mixing as long as the fluid velocity does not exceed a specified value. This indicates that the fluid particles in a glass pipe move in infinite numbers of thin cylindrical layers with one layer gliding smoothly over an adjacent layer. The velocity of fluid particles at the inner wall of the pipe

Velocity distribution

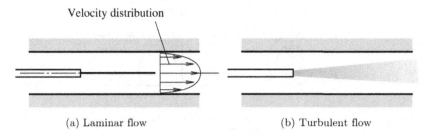

(a) Laminar flow   (b) Turbulent flow

Fig. 2.17   Laminar flow and turbulent flow.

is zero. Such flow is called laminar flow. Increasing the fluid velocity in the pipe progressively, the ink flow tends to deviate from a straight line as it proceeds through the pipe. That is, a transition from laminar flow to turbulent flow takes place. In turbulent flow, the fluid particles mix between layers, and move in irregular paths causing an exchange of momentum from one portion to another. The nature of the flow, i.e. whether a laminar or a turbulent regime is distinguished by Reynolds number Re defined as follows:

$$Re = \frac{u_a d}{\nu},\tag{2.71}$$

where u_a and d are the average fluid velocity in the pipeline and the diameter of the pipeline, respectively, and ν is the kinematic viscosity of the fluid. Osborne Reynolds found that the flow in a pipeline changes from a laminar to turbulent flow when Reynolds number increases to a value between 2000 and 4000, and also that it is always laminar flow when Reynolds number is less than 2000. For the purpose of this study, we will assume that the transition from laminar to turbulent flow occurs at the Reynolds number approximately 2300, regardless of the conditions disturbing the flow such as a roughness of the pipe's inner face. The Reynolds number 2300 is referred to as the critical Reynolds number.

Let us now consider the pressure losses of the steady laminar flow in a pipeline with the diameter $d = 2r$ and the length ℓ. As shown in Fig. 2.18, the fluid element with a radius $r - y$ coinciding with the center of the pipeline is dynamically kept in balance in the steady flow by the action of the hydraulic forces due to the pressures, p_1, p_2 and the shear stress τ of the fluids surrounding it. Therefore,

$$\pi(r - y)^2(p_1 - p_2) = 2\pi(r - y)\ell\tau.\tag{2.72}$$

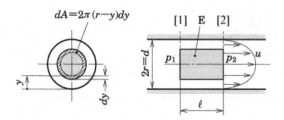

Fig. 2.18   Flow in a pipeline.

Substituting Eq. (2.5) into Eq. (2.72) gives:

$$\frac{(r-y)(p_1-p_2)}{2\ell} = \mu\frac{du}{dy}. \tag{2.73}$$

Regarding the pressures p_1 and p_2 as constant in the radial direction, integrating Eq. (2.73) with respect to the distance y, with the boundary condition $u = 0$ at $y = 0$ gives

$$u = \frac{(2ry-y^2)(p_1-p_2)}{4\mu\ell}. \tag{2.74}$$

Since a flow rate dQ passing through a minute annular ring area dA shown in Fig. 2.18 is described by $dQ = 2\pi u(r-y)dy$, the flow rate Q through the cross-section of the pipe is given as follows:

$$Q = \int_0^r 2\pi u(r-y)dy = \frac{\pi r^4(p_1-p_2)}{8\mu\ell} = \frac{\pi d^4(p_1-p_2)}{128\mu\ell}, \tag{2.75}$$

where d is the diameter of the pipe and ℓ is the length of the pipeline. Equation (2.75) is referred to as the Hagen–Poiseuille equation, and it expresses the relationship between the flow rate Q and the pressure loss $p_1 - p_2$ between sections [1] and [2] for the laminar flow in the pipeline. Recalling the average flow velocity u_a in the pipeline, the flow rate Q is described as follows:

$$Q = \frac{\pi}{4}d^2 u_a. \tag{2.76}$$

Substituting Eq. (2.76) into Eq. (2.75) gives,

$$\Delta p = p_1 - p_2 = \lambda\frac{\ell}{d}\frac{\rho u_a^2}{2}, \tag{2.77}$$

$$\lambda = \frac{64}{Re}, \tag{2.78}$$

where λ is the pipe friction factor that coincides with the experimental results for the Reynolds numbers not exceeding 2000, regardless of the roughness on the inner face of the pipeline.

Turning now to turbulent flow, the pipe friction factor λ depends not only on Reynolds number but also on the roughness of the pipe's inner walls. For turbulent flow in the types of pipeline generally used in hydraulic systems, the pipe friction factor may be estimated using the following empirical formulas:

(a) *Blasius' formula*

$$4 \times 10^3 \le Re \le 10^5 : \lambda = \frac{0.316}{Re^{\frac{1}{4}}}. \tag{2.79}$$

(b) *Nikuradse's formula*

$$1 \times 10^5 \le Re \le 1 \times 10^8 : \lambda = 0.0032 + 0.221 Re^{-0.237}. \tag{2.80}$$

2.3.4 *Minor Losses*

The pressure losses due to changes of the pipe cross-section and the presence of bends, elbows, joints, valves, etc. in the pipeline are called minor losses. The minor loss Δp is expressed by the following equation using the average flow velocity u_a in the pipeline cross-section and the loss coefficient ζ:

$$\Delta p = \zeta \frac{1}{2} \rho u_a^2. \tag{2.81}$$

2.3.4.1 *Loss Due to Sudden Expansion of a Conduit*

As shown in Fig. 2.19, the sudden expansion from the cross-sectional area a_1 to a_2 in the pipeline produces the pressure drop Δp between the cross-section [1] and [2]. Applying Eq. (2.60) to the fluid at the cross-section [1] and [2] gives

$$\frac{1}{2}\rho u_1^2 + p_1 = \frac{1}{2}\rho u_2^2 + p_2 + \Delta p. \tag{2.82}$$

Therefore:

$$\Delta p = \frac{1}{2}\rho(u_1^2 - u_2^2) + p_1 - p_2. \tag{2.83}$$

Let us consider the region indicated by broken line shown in Fig. 2.19 as a control volume in order to derive the loss coefficient ζ theoretically.

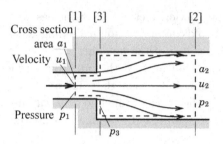

Fig. 2.19   Sudden expansion in a pipeline.

Applying the momentum theory Eq. (2.55) to the fluid in the control volume gives the steady force F_x acting axially on the fluid in the control volume:

$$F_x = \rho a_2 u_2^2 - \rho a_1 u_1^2. \tag{2.84}$$

Since the pressure p_1 at the section [1] is regarded as equal to pressure p_3 at the section [3], the force F_x is obtained by considering the hydraulic force acting axially on the fluid in the control volume as follows:

$$F_x \approx a_2(p_1 - p_2). \tag{2.85}$$

Using Eqs. (2.84) and (2.85), the pressure difference $\Delta p = p_1 - p_2$ is expressed as follows:

$$p_1 - p_2 = \rho\left(u_2^2 - \frac{a_1}{a_2}u_1^2\right). \tag{2.86}$$

By virtue of the principle of continuity given by Eq. (2.42), the relationship between the mean velocities u_1 and u_2 may be described as follows:

$$u_2 = \frac{a_1}{a_2}u_1. \tag{2.87}$$

Where a_1 and a_2 are the cross-section areas of the flow path at the locations [1] and [2], respectively. Substituting Eqs. (2.86) and (2.87) into Eq. (2.83) gives

$$\Delta p = \left(1 - \frac{a_1}{a_2}\right)^2 \frac{1}{2}\rho u_1^2. \tag{2.88}$$

Substituting Eq. (2.81) into Eq. (2.88) gives the loss coefficient as follows:

$$\zeta = \left(1 - \frac{a_1}{a_2}\right)^2. \tag{2.89}$$

It is known that Eq. (2.89) agrees well with experimental results.

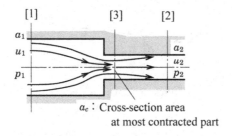

Fig. 2.20 Sudden contraction in a pipeline.

Table 2.2 Contraction factor $C_c = a_c/a_2$.

a_2/a_1	0.1	0.3	0.5	0.7	0.9
C_c	0.632	0.643	0.681	0.755	0.892

2.3.4.2 *Loss Due to Sudden Contraction of a Conduit*

The flow in the pipeline contracts rapidly as it approaches a sudden reduction in the diameter of the conduit, and the cross-section area of the flow path becomes the minimum just downstream of the contraction inlet as shown by the location [3] in Fig. 2.20. Then, the flow path grows gradually to reach the full cross-section area of the conduit.

Pressure is lost in the expanding flow region from the location [3] to [2] in Fig. 2.20, whereas it is negligibly small in the flow contraction region from the location [1] to [3]:

The pressure loss due to a sudden contraction of the conduit can be obtained by applying Eq. (2.89) to the flow region expanding from the cross-section [3] to [2]:

$$\Delta p = \zeta \frac{1}{2}\rho u_2^2, \quad \zeta = \left(\frac{a_2}{a_c} - 1\right)^2 = \left(\frac{1}{C_c} - 1\right)^2, \qquad (2.90)$$

where a_2 and a_c denote the cross-section areas of the pipeline at the locations [2] and [3], respectively as shown in Fig. 2.20. Table 2.2 gives the values of the contraction factor $C_c = a_c/a_2$ calculated for a range of values of the ratio of the outlet area to the inlet area, a_2/a_1.

2.3.4.3 *Pressure Loss Due to Gradual Expansion*

The pressure loss in the gradual expansion flow as shown in Fig. 2.21 has been investigated by experimental methods. The results are summarized as

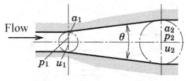

p : Pressure   a : Cross-section
u : Velocity   θ : Angle

Fig. 2.21   Flow in a gradually expanding pipeline.

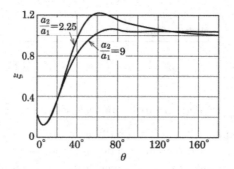

Fig. 2.22   Experimental values of coefficient ξ.

follows:

$$\Delta p = \zeta \frac{1}{2} \rho u_1^2, \quad \zeta = \xi \left[1 - \left(\frac{a_1}{a_2} \right) \right]^2. \tag{2.91}$$

Values of the coefficient ξ in Eq. (2.91) have been investigated experimentally. A selection of results is shown in Fig. 2.22. It may be seen that the coefficient has a minimum value of 0.135 at $\theta = 5$–6.5°. In gradually contracting flow, losses are negligible, and the loss coefficient ζ in Eq. (2.91) may be taken to be zero.

2.3.4.4 *Pressure Loss in Devices Connecting the Pipelines*

As shown in Eq. (2.92), it may be necessary to account for pressure losses Δp through such features as bends, elbows, valves, etc., in a pipeline with the average velocity u_a. Table 2.3 shows the approximate relevant values of the coefficient ζ for a variety of such features, including the three types of inlet shown in Fig. 2.23.

$$\Delta p = \zeta \frac{1}{2} \rho u_a^2. \tag{2.92}$$

Table 2.3 Loss coefficient ζ for devices connecting the pipeline.

Devices		ζ
Check valve (full open)		2.5
Angle valve (full open)		5.0
Globe valve (full open)		10.0
Elbow		1.8
Standard tee		0.9
Entrance loss	Bell mouth	0.05
	Sharp edge	0.5
	Projection	1.0

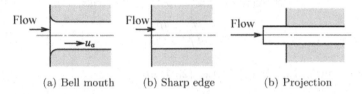

(a) Bell mouth (b) Sharp edge (b) Projection

Fig. 2.23 Types of inlet.

2.3.5 *Flow in a Clearance*

Since most hydraulic components have moving parts such as spools, plungers, vanes and gears, etc., there must be clearance flows between them and the walls of the hydraulic components. The leakage flow in a clearance may cause the performance of the component to deteriorate because the fluid force acts on the moving parts. This is true even when the leakage is small.

In this section, the flow in clearances will be explained. It will be assumed that the flow is both steady and laminar, i.e. that Reynolds number is less than 2000.

2.3.5.1 *Flow in a Clearance Between Two Flat Boards Configured in Parallel*

A fluid flows steadily in a clearance between a fixed flat board and a moving flat board with the velocity U. As shown in Fig. 2.24, considering the minute fluid element E with unit width for the vertical direction on x–y plane, it is kept in a state of dynamic balance by the pressures $p, p + dp$ and the shear stresses τ and $\tau + d\tau$. Then the dynamical balancing equation for the

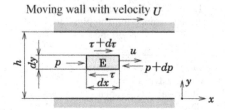

Fig. 2.24   Flow in the clearance between two boards in parallel.

element E with the volume $dxdydz$ is described as follows:

$$pdy + (\tau + d\tau)dx = (p + dp)dy + \tau dx, \qquad (2.93)$$

$$\therefore \frac{d\tau}{dy} = \frac{dp}{dx}. \qquad (2.94)$$

Substituting Eq. (2.5) into Eq. (2.94) gives

$$\frac{d^2u}{dy^2} = \frac{1}{\mu}\frac{dp}{dx}. \qquad (2.95)$$

The velocity u may be derived by twice integrating Eq. (2.95) with respect to y, because it may be assumed that the pressure p remains constant with respect to y. Thus,

$$u = \frac{1}{\mu}\frac{dp}{dx}\frac{y^2}{2} + C_1 y + C_2. \qquad (2.96)$$

Substituting the boundary conditions, i.e. $u = 0$ at $y = 0$ and $u = U$ at $y = h$ into Eq. (2.96), the integration constants C_1 and C_2 are obtained as

$$C_1 = \frac{U}{h} - \frac{1}{\mu}\frac{dp}{dx}\frac{h}{2}, \quad C_2 = 0, \qquad (2.97)$$

Therefore, the velocity distribution in the y-direction is described as follows:

$$u = \frac{U}{h}y - \frac{1}{2\mu}\frac{dp}{dx}(h - y)y. \qquad (2.98)$$

Such flow is referred to Couette flow, and it is often recalled when analyzing clearance flows in hydraulic components. The flow rate q per unit width

$(dz = 1)$ normal to the illustration in Fig. 2.24 is described as

$$q = \int_0^h u\,dy = \frac{U}{2}h - \frac{1}{12\mu}\left(\frac{dp}{dx}\right)h^3. \tag{2.99}$$

Equation (2.99) may be arranged as follows:

$$\frac{dp}{dx} = \frac{12\mu}{h^3}\left(\frac{Uh}{2} - q\right). \tag{2.100}$$

Equation (2.100) shows that the pressure gradient dp/dx is constant, because the flow rate q is constant under the steady flow conditions, and it may be expressed as

$$-\frac{dp}{dx} = \frac{p_1 - p_2}{\ell}, \tag{2.101}$$

where $(p_1 - p_2)$ is the pressure difference and ℓ is the length of the clearance. Substituting $U = 0$ and Eq. (2.101) into Eqs. (2.98) and (2.100) respectively, the fluid velocity u and the flow rate q are obtained as follows:

$$u = \frac{1}{2\mu}\frac{p_1 - p_2}{\ell}(h - y)y, \tag{2.102}$$

$$q = \frac{h^3}{12\mu}\frac{p_1 - p_2}{\ell}. \tag{2.103}$$

2.3.5.2 *Flow in an Annular Clearance*

Flow in an annular clearance may occur between a spool and the bore wall surrounding it in a hydraulic component. When the fluid flows through the concentric annular clearance between the bore with the diameter $2R$ and the shaft with the diameter $2r$ as shown in Fig. 2.25(a), it may be regarded as the flow in the parallel clearance between two flat boards with length ℓ and the width $2\pi R$. Therefore, multiplying Eq. (2.103) by $2\pi R$ gives the flow rate Q_0 through the annular clearance:

$$Q_0 = \frac{\pi R h^3}{6\mu}\frac{p_1 - p_2}{\ell}. \tag{2.104}$$

Let us consider the fluid flow through the infinitesimal area $R\,d\theta\,dy$ at an angle θ and a distance y in the eccentric annular clearance, as shown in Fig. 2.25(b). Since the fluid velocity u is defined by Eq. (2.102), the flow

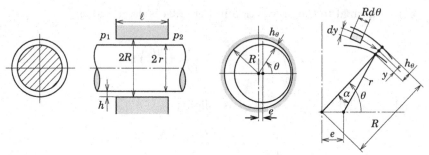

(a) Concentric annular clearance　　　　(b) Eccentric annular clearance

Fig. 2.25　Flow in an annular clearance.

rate dQ through the minute area $Rd\theta h_\theta$ is described as follows:

$$dQ = Rd\theta \int_0^{h_\theta} udy = Rd\theta \int_0^{h_\theta} \frac{1}{2\mu}\frac{p_1-p_2}{\ell}y\,(h_\theta - y)dy = \frac{Rd\theta}{12\mu}\frac{p_1-p_2}{\ell}h_\theta^3.$$

(2.105)

Using notations of the angles α, θ and the eccentricity distance e as shown in Fig. 2.25(b), the height of the clearance h_θ is given as

$$h_\theta = R - r\cos\alpha - e\cos\theta \approx R - r - e\cos\theta = h - e\cos\theta. \qquad (2.106)$$

Substituting Eq. (2.106) into Eq. (2.105), the flow rate Q through the eccentric annular clearance is given as follows:

$$Q = \int_0^{2\pi} \frac{R}{12\mu}\frac{p_1-p_2}{\ell}(h - e\cos\theta)^3 d\theta. \qquad (2.107)$$

To solve Eq. (2.107), we use the definite integral formulas:

$$\left.\begin{array}{l} \displaystyle\int_0^{2\pi}\cos^2\theta d\theta = \left[\frac{\theta}{2} + \frac{1}{4}\sin 2\theta\right]_0^{2\pi} = \pi \\[4mm] \displaystyle\int_0^{2\pi}\cos^3\theta d\theta = \left[\frac{\sin 3\theta}{12} + \frac{3}{4}\sin\theta\right]_0^{2\pi} = 0 \end{array}\right\}. \qquad (2.108)$$

Thus, the flow rate Q is derived as follows:

$$Q = \frac{\pi R h^3}{6\mu}\frac{(p_1-p_2)}{\ell}\left(1 + \frac{3e^2}{2h^2}\right). \qquad (2.109)$$

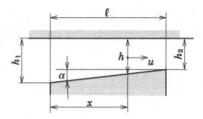

Fig. 2.26   Flow in an inclined clearance between plain stationary walls.

As shown in Eqs. (2.104) and (2.109), the flow rate Q through an annular clearance with the eccentricity rate e/h corresponds to $1+(3/2)(e/h)^2$ times of the flow rate through the concentric annular clearance.

2.3.5.3 *Flow in a Contracting or Expanding Clearance*

(a) *Flow in a clearance surrounded by stationary walls*

Equations (2.98) to (2.103) may be applicable to a clearance h that varies slightly with the distance x, as shown in Fig. 2.26. Substituting velocity $U = 0$ into Eq. (2.99) gives

$$q = \frac{-h^3}{12\mu}\left(\frac{dp}{dx}\right). \tag{2.110}$$

When the board has the small inclination $\alpha \approx \tan\alpha$ against the other board, as shown in Fig. 2.26, the clearance h varies with distance x:

$$\left. \begin{aligned} h &= h_1 - \alpha x \\ \alpha &= \frac{(h_1 - h_2)}{\ell} \end{aligned} \right\}. \tag{2.111}$$

Substituting Eq. (2.111) into Eq. (2.110) gives

$$p = \int \frac{-12\mu q}{(h_1 - \alpha x)^3} dx = -\frac{6\mu q}{\alpha(h_1 - \alpha x)^2} + C. \tag{2.112}$$

Canceling the integration constant C from the two equations obtained by substituting the boundary conditions (i.e. $p = p_1$ at $x = 0$ and $p = p_2$ at $x = \ell$) into Eq. (2.112), yields the following equation:

$$q = \frac{1}{6\mu\ell} \frac{(h_1 h_2)^2}{(h_1 + h_2)}(p_1 - p_2). \tag{2.113}$$

Substituting the boundary condition $p = p_1$ at $x = 0$, into Eq. (2.112) gives the integral constant as $C = [6\mu q/(\alpha h_1)^2] + p_1$, and the pressure p at the clearance h is obtained as follows:

$$p = -\frac{6\mu q}{\alpha}\frac{h_1^2 - h^2}{h_1^2 h^2} + p_1. \tag{2.114}$$

Substituting Eq. (2.113) into Eq. (2.114) gives

$$p = p_1 - \frac{\left(\frac{h_1}{h}\right)^2 - 1}{\left(\frac{h_1}{h_2}\right)^2 - 1}(p_1 - p_2). \tag{2.115}$$

Equations (2.111) to (2.115) are applicable to both flows in the expanding clearance ($h_1 < h_2$) and in the contracting clearance ($h_1 > h_2$). Let us consider the flows in both types of clearances by considering the upper and lower surface of a taper plate placed between two parallel flat boards, with an eccentricity distance e, as shown in Fig. 2.27. The inlet and outlet upper clearances h_{U1}, h_{U2} and the inlet and outlet lower clearances h_{L1}, h_{L2} are denoted as follows:

$$\left.\begin{array}{ll} h_{U1} = h_{10} - e, & h_{U2} = h_{20} - e \\ h_{L1} = h_{10} + e, & h_{L2} = h_{20} + e \end{array}\right\}, \tag{2.116}$$

where h_{10} and h_{20} are respectively the inlet and the outlet clearance under the condition that the eccentric distance e is zero. Then, the upper clearance h_U and the lower clearance h_L at the distance x are described by using

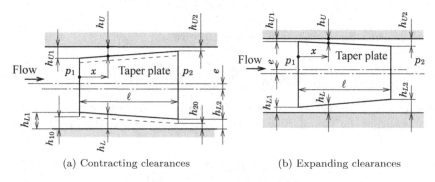

(a) Contracting clearances   (b) Expanding clearances

Fig. 2.27   Flows in the clearances above and below a taper plate.

Eq. (2.111) and (2.116) as follows:

$$h_U = h_{10} - e - \frac{(h_{10} - h_{20})x}{\ell}, \tag{2.117}$$

$$h_L = h_{10} + e - \frac{(h_{10} - h_{20})x}{\ell}. \tag{2.118}$$

The clearance h_U or h_L and the pressure p_U or p_L in Fig. 2.27 correspond to the clearance h and pressure p in Eq. (2.115), respectively. Accordingly, recalling Eq. (2.115) to (2.118), the pressure difference $\Delta p = p_L - p_U$ between the upper and lower clearances in Fig. 2.27 may be derived as follows:

$$\Delta p = p_L - p_U = \left[\frac{\left(\frac{h_{10}-e}{h_U}\right)^2 - 1}{\left(\frac{h_{10}-e}{h_{20}-e}\right)^2 - 1} - \frac{\left(\frac{h_{10}+e}{h_L}\right)^2 - 1}{\left(\frac{h_{10}+e}{h_{20}+e}\right)^2 - 1} \right] (p_1 - p_2). \tag{2.119}$$

Using h_{10} and ℓ as the reference length, the dimensionless parameters are defined as follows:

$$\bar{e} = \frac{e}{h_{10}}, \quad \bar{h}_{20} = \frac{h_{20}}{h_{10}}, \quad \bar{x} = \frac{x}{\ell}. \tag{2.120}$$

Thus, Eq. (2.119) may be rewritten in a dimensionless format:

$$\Delta\bar{p} = \frac{p_L - p_U}{p_1 - p_2} = \frac{\left[\frac{1-\bar{e}}{1-\bar{e}-(1-\bar{h}_{20})\bar{x}}\right]^2 - 1}{\left(\frac{1-\bar{e}}{\bar{h}_{20}-\bar{e}}\right)^2 - 1} - \frac{\left[\frac{1+\bar{e}}{1+\bar{e}-(1-\bar{h}_{20})\bar{x}}\right]^2 - 1}{\left(\frac{1+\bar{e}}{\bar{h}_{20}+\bar{e}}\right)^2 - 1}. \tag{2.121}$$

Figure 2.28 shows the variation of the dimensionless pressure difference $\Delta\bar{p}$ between the lower and upper clearances.

In the expanding clearance flow as shown in Fig. 2.27(b), the dimensionless pressure difference $\Delta\bar{p}$ has positive values, and therefore the hydraulic force acts on the taper plate in the direction whereby it is pressed toward the upper wall. On the other hand, the pressure difference $\Delta\bar{p}$ in the contracting clearance flow shown Fig. 2.27(a) has negative values, so the hydraulic force acts on the taper plate such that the eccentric distance e is decreased, provided the taper plate position remains unchanged.

Figure 2.29 illustrates the flow in a clearance between a bore and a taper spool located with an eccentric distance e. The axial flows in the set of expanding taper spool clearances located at θ and $\theta + \pi$ in Fig. 2.29 correspond to the flows in the set of the upper and lower clearances in Fig. 2.27(b) though the taper spool clearances tend to vary nonlinearly

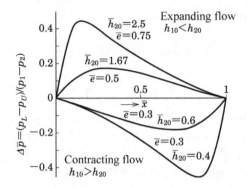

Fig. 2.28   Distribution of the dimensionless pressure difference.

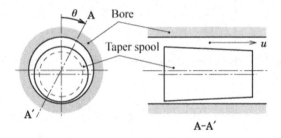

Fig. 2.29   Clearance between a bore and a tapered shaft.

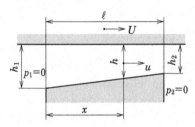

Fig. 2.30   Flow in contracting clearance with a moving wall.

with the angle θ. Therefore, the flow in the expanding annular taper spool clearance produces a hydraulic force pressing the spool toward the bore, and the force causes the spool to become locked. This phenomenon experienced in hydraulic systems is referred to as hydraulic lock.

(b) *Flow due to a moving wall in a contracting clearance*

Figure 2.30 illustrates a two-dimensional flow in the clearance between a fixed taper board and a moving flat board, with the velocity U. When the flat board is moved to the right, the fluids are dragged into the clearance between the boards due to the fluid viscosity, and then the pressure p in the clearance increases, while the inlet and outlet pressures remain at atmospheric pressure. Substituting Eq. (2.111) into Eq. (2.100) gives

$$\frac{dp}{dx} = \frac{6\mu U}{(h_1 - \alpha x)^2} - \frac{12\mu q}{(h_1 - \alpha x)^3}. \tag{2.122}$$

Integrating Eq. (2.122) with respect to x gives the pressure p at the distance x:

$$p = \int \left[\frac{6\mu U}{(h_1 - \alpha x)^2} - \frac{12\mu q}{(h_1 - \alpha x)^3} \right] dx = \frac{6\mu U}{\alpha(h_1 - \alpha x)} - \frac{6\mu q}{\alpha(h_1 - \alpha x)^2} + C. \tag{2.123}$$

Substituting the boundary conditions (i.e. $p = 0$ at $x = 0$ and $p = 0$ at $x = \ell$) into Eq. (2.123) gives

$$q = \frac{h_1 h_2}{h_1 + h_2} U. \tag{2.124}$$

$$C = \frac{-6\mu U}{\alpha (h_1 + h_2)}. \tag{2.125}$$

Substituting Eqs. (2.124) and (2.125) into Eq. (2.123) gives

$$p = \frac{6\mu U (h_1 - h_2 - \alpha x) x}{(h_1 + h_2)(h_1 - \alpha x)^2}. \tag{2.126}$$

The hydraulic force F acting on the clearance wall per unit width is described as:

$$F = \int_0^\ell p dx = \frac{6\mu U \ell^2}{(h_1 - h_2)^2} \left(\ln\frac{h_1}{h_2} - 2\frac{h_1 - h_2}{h_1 + h_2} \right). \tag{2.127}$$

Here, considering the following dimensionless parameter:

$$\bar{U} = \frac{U}{\ell}, \quad \bar{\ell} = \frac{\ell}{h_1}, \quad \bar{h}_2 = \frac{h_2}{h_1}, \quad \bar{x} = \frac{x}{\ell}. \tag{2.128}$$

Equation (2.126) may be arranged in a dimensionless form using the above parameters:

$$\bar{p} = \frac{p}{6\mu \bar{U} \bar{\ell}^2} = \frac{[1 - \bar{h}_2 - (1 - \bar{h}_2)\bar{x}]\bar{x}}{(1 + \bar{h}_2)[1 - (1 - \bar{h}_2)\bar{x}]^2}. \tag{2.129}$$

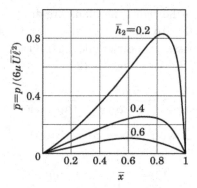

Fig. 2.31   Pressure distribution in a contracting clearance with a moving wall.

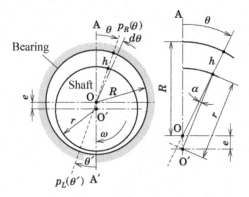

Fig. 2.32   Journal bearing.

Figure 2.31 shows dimensionless pressure distributions computed using Eq. (2.129).

2.3.5.4 *Journal Bearings*

Figure 2.32 shows a journal bearing that supports the load on a rotating shaft by means of the hydraulic pressure occurring in the clearance between the bore and shaft. The shaft rotation produces a flow in the contracting clearance at the right side of the reference line a–a' shown in Fig. 2.32, which generates a flow in the expanding clearance at the left side. The contracting clearance h at an angle θ is defined as follows:

$$h = R - r\cos\alpha + e\cos\theta \approx h_0 + e\cos\theta, \quad (0 \leq \theta \leq \pi), \qquad (2.130)$$

where

$$h_0 = R - r, \tag{2.131}$$

while the expanding clearance h at an angle θ' is given as

$$h = R - r \cos\alpha - e \cos\theta' \approx h_0 - e \cos\theta', \quad (0 \le \theta' \le \pi). \tag{2.132}$$

Applying Eq. (2.100) to the flow in the eccentric annular clearance in Fig. 2.32, the wall velocity U and the distance x may be used as $R\omega$, $R\theta$ (or $R\theta'$), respectively, because the mean clearance $R - r$ is very small in comparison with the shaft radius r. Then, substituting Eq. (2.130) into Eq. (2.100) gives the pressure p_R in the annular contracting clearance:

$$p_R = R \int \frac{6\mu R\omega}{(h_0 + e\cos\theta)^2} d\theta - R \int \frac{12\mu q}{(h_0 + e\cos\theta)^3} d\theta. \tag{2.133}$$

Substituting Eq. (2.132) into Eq. (2.100) gives the pressure p_L in the annular expansion clearance.

$$p_L = R \int \frac{6\mu R\omega}{(h_0 - e\cos\theta')^2} d\theta' - R \int \frac{12\mu q}{(h_0 - e\cos\theta')^3} d\theta'. \tag{2.134}$$

The pressures p_R and p_L can be derived as the respective functions $f_R(\theta)$ and $f_L(\theta')$, respectively, using the following integral formulas:

$$\int \frac{d\theta}{(h_0 \pm e\cos\theta)^2}$$

$$= \frac{1}{h_0^2 - e^2} \left[-\frac{\pm e \sin\theta}{h_0 \pm e\cos\theta} + \frac{2h_0}{\sqrt{h_0^2 - e^2}} \tan^{-1} \frac{\sqrt{h_0^2 - e^2} \tan\left(\frac{\theta}{2}\right)}{h_0 \pm e} \right]. \tag{2.135}$$

$$\int \frac{d\theta}{(h_0 \pm e\cos\theta)^3}$$

$$= \frac{1}{2(h_0^2 - e^2)} \left[-\frac{\pm e \sin\theta}{(h_0 \pm e\cos\theta)^2} - \frac{\pm 3eh_0 \sin\theta}{(h_0^2 - e^2)(h_0 \pm e\cos\theta)} \right]$$

$$+ \frac{2h_0^2 + e^2}{(h_0^2 - e^2)^2 \sqrt{h_0^2 - e^2}} \tan^{-1} \frac{\sqrt{h_0^2 - e^2} \tan\left(\frac{\theta}{2}\right)}{h_0 \pm e}. \tag{2.136}$$

The pressures p_R and p_L are defined as follows:

$$p_R(\theta) = f_R(\theta) + C_1 \quad (0 \leq \theta \leq \pi), \tag{2.137}$$

$$p_L(\theta') = f_L(\theta') + C_2 \quad (0 \leq \theta' \leq \pi), \tag{2.138}$$

where C_1 and C_2 are integral constants. Recalling Equations (2.133) to (2.138), the values of functions $f_R(0), f_L(0), f_R(\pi)$ and $f_L(\pi)$ are derived as follows:

$$f_R(0) = f_L(0) = 0, \tag{2.139}$$

$$f_R(\pi) = f_L(\pi) = \frac{6\pi\mu R^2 \omega h_0}{(h_0^2 - e^2)\sqrt{h_0^2 - e^2}} - \frac{6\pi\mu Rq(2h_0^2 + e^2)}{(h_0^2 - e^2)^2\sqrt{h_0^2 - e^2}}. \tag{2.140}$$

The boundary conditions $p_R(0) = p_L(\pi), p_R(\pi) = p_L(0)$ between the contracting and the expanding clearance yield the following equations:

$$C_1 = \frac{6\pi\mu R^2 \omega h_0}{(h_0^2 - e^2)\sqrt{h_0^2 - e^2}} - \frac{6\pi\mu Rq(2h_0^2 + e^2)}{(h_0^2 - e^2)^2\sqrt{h_0^2 - e^2}} + C_2, \tag{2.141}$$

$$C_2 = \frac{6\pi\mu R^2 \omega h_0}{(h_0^2 - e^2)\sqrt{h_0^2 - e^2}} - \frac{6\pi\mu Rq(2h_0^2 + e^2)}{(h_0^2 - e^2)^2\sqrt{h_0^2 - e^2}} + C_1. \tag{2.142}$$

Therefore,

$$C_1 = C_2, \quad q = \frac{h_0 R\omega(h_0^2 - e^2)}{2h_0^2 + e^2}. \tag{2.143}$$

The pressure difference $\Delta p = p_R(\theta) - p_L(\theta')$ expresses the pressure difference between the contracting and the expanding annular clearance for the radial direction through the bore center. Recalling Eqs. (2.133), (2.134) and (2.143), the pressure difference $\Delta p = p_R(\theta) - p_L(\theta')$ can be defined as follows:

$$\Delta p = p_R(\theta) - p_L(\theta')$$

$$= 6\mu R^2 \omega \int_0^\theta \left[\frac{1}{(h_0 + e\cos\theta)^2} - \frac{1}{(h_0 - e\cos\theta)^2} \right] d\theta$$

$$- \frac{12\mu h_0 R^2 \omega(h_0^2 - e^2)}{2h_0^2 + e^2} \int_0^\theta \left[\frac{1}{(h_0 + e\cos\theta)^3} - \frac{1}{(h_0 - e\cos\theta)^3} \right] d\theta. \tag{2.144}$$

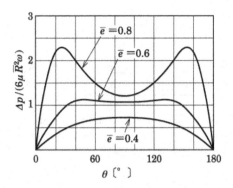

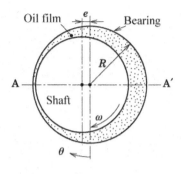

Fig. 2.33   Pressure difference distributions.

Substituting Eqs. (2.135) and (2.136) into Eq. (2.144) gives

$$\frac{\Delta p}{6\mu \bar{R}^2 \omega} = \frac{4\bar{e}\sin\theta}{(2+\bar{e}^2)(1-\bar{e}^2\cos^2\theta)^2},\qquad(2.145)$$

where $\bar{e} = e/h_0$ $\bar{R} = R/h_0$. Figure 2.33 shows the dimensionless pressure distributions of $\Delta\bar{p} = \Delta p/(6\mu\bar{R}^2\omega)$ derived using Eq. (2.145).

The differential pressure distributions $\Delta\bar{p}$ are derived on the assumption that the eccentric distance e remains constant. However, the eccentricity actually does fluctuate because of the actual force balance between the shaft load and the force due to the pressure experienced in the clearance.

2.3.5.5 *Hydrostatic Bearings*

Figure 2.34 illustrates a hydrostatic bearing into which hydraulic fluids are supplied through a center hole with the radius r_1. The fluid flows radially and it is discharged at atmospheric pressure. Then, the load on the upper disk with radius r_2 is supported by the pressure occurring in the clearance of the hydrostatic bearing.

Let us consider an annular fluid element with an infinitesimal cross-section area $dydr$ at a radius r in the clearance, as shown in Fig. 2.34. The shear stresses τ and $\tau + d\tau$ act on the lower and the upper surface of the element, and the pressures p and $p + dp$ act on the inner and the outer ring surface of the element, respectively. Accordingly, the hydraulic forces acting on the element dynamically balanced in the steady flows can be given as

$$2p\pi r dy + 2(\tau+d\tau)\pi r dr = 2(p+dp)\pi(r+dr)dy + 2\tau\pi r dr.\qquad(2.146)$$

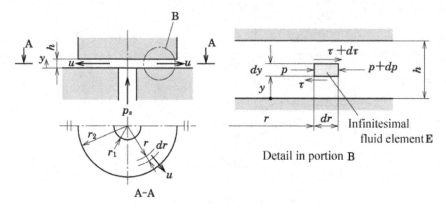

Fig. 2.34   Hydrostatic bearing.

The infinitesimal term dr/r being omitted, Eq. (2.146) can be arranged as follows:

$$\frac{d\tau}{dy} \approx \frac{dp}{dr}. \tag{2.147}$$

Substituting Eq. (2.5) into Eq. (2.147) gives

$$\frac{d^2u}{dy^2} = \frac{1}{\mu}\frac{dp}{dr}, \tag{2.148}$$

where u is the velocity and it is assumed that the pressure p keeps constant regardless of the distance y.

Twice integrating Eq. (2.148) with respect to y gives the velocity distribution equation with integral constants, and then the integral constants are derived by substituting the boundary conditions $u = 0$ at $y = 0$, $u = 0$ at $y = h$ into the velocity distribution equation. Accordingly, we get

$$u = -\frac{1}{2\mu}\frac{dp}{dr}y(h - y). \tag{2.149}$$

Since the fluid moving with velocity u passes through the infinitesimal annular cross-section area $2\pi r dy$, the flow rate Q through the clearance is expressed as follows:

$$Q = \int_0^h 2\pi r u\, dy = -\frac{\pi r}{\mu}\frac{dp}{dr}\int_0^h y(h - y)\, dy = -\frac{\pi r h^3}{6\mu}\frac{dp}{dr}, \tag{2.150}$$

where the flow rate Q is constant in the steady flow. The relationship between a radius r and pressure p is expressed as follows, recalling

Eq. (2.150):

$$-\int_{r_1}^{r} \frac{1}{r} dr = \int_{p_s}^{p} \frac{\pi h^3}{6\mu Q} dp,$$ (2.151)

$$\ln \frac{r_1}{r} = \frac{\pi h^3 (p - p_s)}{6\mu Q}.$$ (2.152)

Substituting the boundary condition $p = 0$ at $r = r_2$ into Eq. (2.152) gives

$$Q = \frac{\pi h^3 p_s}{6\mu \ln \left(\frac{r_2}{r_1}\right)}.$$ (2.153)

Substituting Eq. (2.153) into Eq. (2.152) gives the dimensionless pressure term:

$$\frac{p}{p_s} = 1 + \frac{\ln \left(\frac{r_1}{r}\right)}{\ln \left(\frac{r_2}{r_1}\right)}.$$ (2.154)

Using Eq. (2.154), the thrust force F due to the pressure in the hydrostatic bearing is defined as follows:

$$F = \pi r_1^2 p_s + \int_{r_1}^{r_2} 2\pi r p \, dr = \frac{\pi r_1^2 p_s}{2 \ln \left(\frac{r_2}{r_1}\right)} \left[\left(\frac{r_2}{r_1}\right)^2 - 1\right].$$ (2.155)

Figure 2.35 illustrates the dimensionless pressure distribution, and Fig. 2.36 shows the relationship between the dimensionless thrust forces $F/(\pi r_1^2 p_s)$ and the dimensionless radius r_2/r_1.

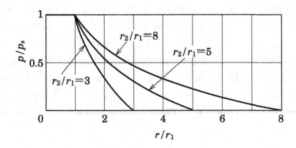

Fig. 2.35   Pressure distribution in a hydrostatic bearing.

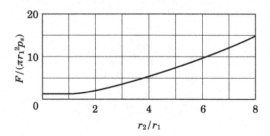

Fig. 2.36   Thrust of a hydrostatic bearing.

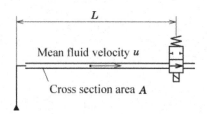

Fig. 2.37   Transmission line in the system.

2.4 Oil Hammer

2.4.1 *Oil Hammer in a Rigid Transmission Line*

At the instant when the fluid flow in a pipeline has stopped suddenly by
shutting a valve, a high pressure pulse is generated, and the surge pressure
propagates upstream as a pressure wave. The action of the pressure pulse
causes the velocity of the flow upstream to decrease and, as a consequence,
the flow stops. On the downstream side of the valve, the pressure is reduced
and the lower pressure wave propagates to the downstream regions. The
phenomenon generating such a surge pressure is known as oil hammer.

Let us consider the pressure rise due to oil hammer. As shown in
Fig. 2.37, we will consider a hydraulic fluid with density ρ flowing at a
velocity u in a pipeline with length L and cross-sectional area A. The pressure rise Δp due to oil hammer propagates through the pipeline to the
upstream section, in the form of the pressure wave. Putting that the pressure propagation velocity i.e. acoustic velocity is a and the pipeline length
is L, the propagating time Δt is described as follows:

$$\Delta t = \frac{L}{a}. \tag{2.156}$$

In the time that the pressure wave travels the distance L upstream, the flow upstream is stopped. Thus, the propagation time Δt corresponds to the time taken to stop the fluid flow in the upstream side of the pipeline. That is, the momentum of the fluid flowing in the pipeline changes from ρALu to zero during the time Δt. The kinetic energy of the fluid is converted into pressure energy. Applying Newton's equation of motion to the fluid flowing in the pipeline, the force acting on the fluid is expressed as

$$\Delta p A = \lim_{\Delta t \to 0} \frac{\rho ALu - 0}{\Delta t}. \tag{2.157}$$

Substituting Eq. (2.156) into above Eq. (2.157) gives

$$\Delta p = \rho u a. \tag{2.158}$$

When the fluid volume in the pipeline changes from AL to $AL + dV$ at the pressure p, the energy required for the fluid volume change becomes $dW_c = -pdV$. Since the pressure rise Δp and volume change ΔV in the pipeline occur due to the oil hammer effect, the compression energy W_c is described as follows:

$$W_c = \int_0^{\Delta V} p(-dV). \tag{2.159}$$

The volume change dV is expressed as $-dV = (AL/\kappa)dp$ using the bulk modulus κ defined by Eq. (2.11). Substituting the volume change $-dV$ into Eq. (2.159) gives

$$W_c = \int_0^{\Delta p} \frac{ALp}{\kappa} dp = \frac{\Delta p^2 AL}{2\kappa}. \tag{2.160}$$

Since the kinetic energy $W_k = \rho ALu^2/2$ of the fluid flowing through the pipeline is converted into the compression energy W_c by the pressure rise Δp due to the oil hammer effect, we obtain the following equation:

$$\frac{\rho ALu^2}{2} = \frac{\Delta p^2 AL}{2\kappa}. \tag{2.161}$$

Therefore,

$$\Delta p = u\sqrt{\rho\kappa}. \tag{2.162}$$

Substituting Eq. (2.158) into Eq. (2.162) gives

$$a = \sqrt{\frac{\kappa}{\rho}}. \tag{2.163}$$

Let us treat the fluid in the pipeline with volume V as the system, hence the system mass ρV will not change due to the volume change. The infinitesimal term in $(\rho + d\rho)(V + dV) = \rho V$ may be neglected, and the following approximate equation is derived accordingly:

$$\frac{-dV}{V} \approx \frac{d\rho}{\rho}. \tag{2.164}$$

Applying Eq. (2.11), the ratio dV/V is expressed as $dV/V = -dp/\kappa$. Substituting dV/V into Eq. (2.164) gives

$$\frac{\kappa}{\rho} \approx \frac{dp}{d\rho}. \tag{2.165}$$

Using Eqs. (2.163) and (2.165), the acoustic velocity a is described as follows:

$$a = \sqrt{\frac{dp}{d\rho}}. \tag{2.166}$$

2.4.2 *Oil Hammer in an Elastic Transmission Line*

When oil hammer occurs in an elastic transmission line, the kinetic energy W_k of the fluid flowing through the pipeline is converted not only into the compression energy of the fluid W_c but also into the energy W_p causing the pipeline volume to enlarge. Thus, we get

$$\frac{\rho A L u^2}{2} = \frac{A L \Delta p^2}{2\kappa} + W_p. \tag{2.167}$$

When the pressure p acts on the inner surface of the pipeline with a diameter d_1, thickness δ and length L as shown in Fig. 2.38, the tensile stress acting on the material of the pipeline is given as follows:

$$\sigma = \frac{d_1}{2\delta} p. \tag{2.168}$$

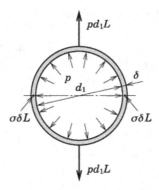

Fig. 2.38 Force and stress acting on a pipe cross-section.

The change of the tensile stress $d\sigma$ resulting from the pressure change dp is given by:

$$d\sigma = \frac{d_1}{2\delta}dp. \qquad (2.169)$$

When the pipe is thin, the circular strain $d\varepsilon$ for the circular elongation ds is given as follows:

$$d\varepsilon = \frac{ds}{\pi d_1}. \qquad (2.170)$$

The modulus of longitudinal elasticity E of the pipe material is defined as follows:

$$E = \frac{d\sigma}{d\varepsilon}. \qquad (2.171)$$

Recalling Eq. (2.169) to (2.171), the circular elongation ds may be expressed as follows:

$$ds = \pi d_1 d\varepsilon = \frac{\pi d_1}{E}d\sigma = \frac{\pi d_1^2}{2E\delta}dp. \qquad (2.172)$$

When the pressure increase Δp due to the oil hammer effect gives rise to circular elongation Δs, the energy W_p involved in enlarging the pipe cross-section area is described as follows:

$$W_p = \int_0^{\Delta s} \sigma \delta L ds. \qquad (2.173)$$

Substituting Eqs. (2.168) and (2.172) into Eq. (2.173) gives

$$W_p = \int_0^{\Delta p} \frac{d_1 p}{2\delta} \frac{\pi d_1^2 L}{2E} dp = \frac{\pi d_1^3 L \Delta p^2}{8E\delta} = \frac{A d_1 L \Delta p^2}{2E\delta}, \quad (2.174)$$

where A is the pipe cross-section area ($A = \pi d_1^2/4$). Substituting Eq. (2.174) into Eq. (2.167) gives

$$\Delta p = \frac{u\sqrt{\rho\kappa}}{\sqrt{1 + \left(\frac{\kappa d_1}{E\delta}\right)}}. \quad (2.175)$$

Substituting Eq. (2.158) into Eq. (2.175) gives

$$a = \frac{\sqrt{\frac{\kappa}{\rho}}}{\sqrt{1 + \left(\frac{\kappa d_1}{E\delta}\right)}}. \quad (2.176)$$

Comparison of Eq. (2.175) with Eq. (2.162) reveals that the effect of pipe elasticity on pressure rise Δp due to oil hammer may be negligible provided that $\kappa d_1/(E\delta) \ll 1$.

One way of limiting the adverse affects of oil hammer is to install an accumulator in the transmission line of the hydraulic system so that the pressure rise due to the oil hammer can be absorbed by the accumulator. The operating principles of the accumulator are explained in Section 3.4.1.

2.5 Cavitation

When the local liquid pressure in the flow falls below saturated vapor pressure, the liquid boils and vapor pockets develop in the liquid. The vapor bubbles travel along with the flow, and suddenly collapse when they enter a higher pressure region. When the bubbles collapse, the liquid rushing into the cavities gives rise to very high local pressures, accompanied by noise. This phenomenon is referred to as cavitation. If the bubbles collapse near the solid boundary, the forces due to the high local pressure give rise to pitting of the solid boundary surface, and are responsible for corrosive wear, vibration of the system and the generation of loud noise.

Although cavitation is caused by the fluid vaporization at very low pressures, a phenomenon known as pseudo-cavitation occurs at pressures significantly higher than the saturated vapor pressure because the hydraulic fluid holds entrapped air which escapes when pressure is locally reduced. In the case of petroleum-based hydraulic fluids, the volumetric entrained air

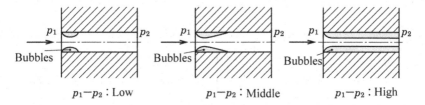

$p_1 - p_2$: Low $p_1 - p_2$: Middle $p_1 - p_2$: High

Fig. 2.39   Cavitation configuration in a cylindrical choke ($p_1 = $ const.).

ratio is from about 7 to 11% under standard atmosphere conditions, and the saturated vapor pressure is about 7 Pa at 20°C. It is known that pseudo-cavitation frequently occurs in a hydraulic system at locations where the local pressures lie in the range 2–8 kPa. This may occur at locations such as the suction port of a pump, the throttle in an orifice or the opening port of a valve in the hydraulic system. It gives rise to undesirable conditions such as loud noise, vibration and damage to flow passages.

For instance, when the rotational speeds of a pump are excessive, small air bubbles may emerge locally in the suction port because of the pressure reduction. Their presence gives rise to an increased flow resistance in the suction port, which impedes the flow. Figure 2.39 shows a configuration where the pseudo-cavitation effect occurs at a cylindrical choke. When pressure p_2 is gradually reduced under the constant upstream pressure p_1, air bubbles are formed at the inlet of the cylindrical choke and grow in the region from the inlet to the outlet of the choke. Thus, we may see that the reduction of the downstream pressure p_2 does not provide the expected increase in flow rate, but instead gives rise to what is known as the cavitation closure effect.

Problems

2.1 A piston with diameter $d = 40$ mm and length $L = 30$ mm moves inside a cylinder at a velocity $u = 1$ m/s. The clearance h between the cylinder and the piston is $15\,\mu$m and it is filled with an oil film which has the viscosity $\mu = 0.04$ Pa·s. Calculate the viscous shearing force required to move the piston.

2.2 As shown in Fig. 2.40, hydraulic oil is enclosed in a cylinder with a length of $L = 400$ mm. Calculate the volume change of the hydraulic oil in the cylinder assuming that a mass $m = 12000$ kg is loaded on the cylinder piston, and the bulk modulus κ of the hydraulic oil is 1.2×10^3 MPa.

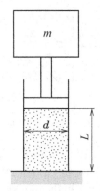

Fig. 2.40   Hydraulic oil enclosed in a cylinder.

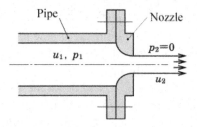

Fig. 2.41   Flow through a nozzle.

2.3 The effective bulk modulus κ_1 of the oil in a steel pipe is 1000 MPa, and the pipe length and the inner diameter are 100 cm and d cm, respectively. The effective bulk modulus κ_2 of the oil in a flexible tube is 860 MPa, and the flexible tube length and the inner diameter are 150 cm and d cm, respectively. Calculate the effective bulk modulus κ_e of the oil in the steel pipe when it is connected to the flexible tube.

2.4 When fluid flows at the rate $Q = 20\,\text{L/min}$ through an orifice in a pipeline, the pressure loss in the orifice-pipeline system is 10 MPa. Calculate the converted thermal power and the fluid temperature increase under the assumption that all of the converted thermal power is expended to increase the fluid temperature. Fluid density ρ is $870\,\text{kg/m}^3$ and the specific heat of the fluid is $c_p = 2.0\,\text{kJ/(kg.°C)}$.

2.5 As shown in Fig. 2.41, hydraulic fluid discharges at atmospheric pressure through a nozzle connected to the pipeline. Calculate the fluid force detaching the nozzle from the pipeline. Pressure p_1 in the

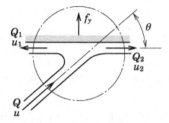

Fig. 2.42   Jet flow impinging on a plate.

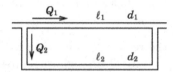

Fig. 2.43   Fluid flows through a two-branch network.

pipeline is $0.43\,\mathrm{MPa}$, the fluid density ρ is $860\,\mathrm{kg/m^3}$, the pipeline diameter d is $12\,\mathrm{cm}$, the nozzle diameter d_n is $3\,\mathrm{cm}$ and the flow coefficient of the nozzle C_d is 0.9.

2.6 As shown in Fig. 2.42, a two-dimensional jet Q impinges on the fixed smooth plate, and divides into flows Q_1 and Q_2. It may be assumed that the losses due to the impingement are negligible.

(1) Find the force f_y acting on the plate in terms of the jet angle θ, the jet velocity u, the flow rate Q and the fluid density ρ.

(2) Find the flow rates Q_1 and Q_2 in terms of the jet angle θ and the flow rate Q.

2.7 Fluid flows through a pipeline with inner diameter d and length ℓ. The pressure losses in the pipeline are denoted by Δp_1 for flow rate Q_1 and Δp_2 for flow rate Q_2. Expressing Q_2 as nQ_1, derive an expression for $\Delta p_1/\Delta p_2$ in each of the following conditions:

(1) Blasius' formula is applicable for the both flow rates.

(2) Both flows are laminar.

2.8 As shown in Fig. 2.43, fluid flows through a two-branch network that consists of a pipeline with inner diameter d_1 and length ℓ_1, and a pipeline with inner diameter d_2 and length ℓ_2. The Reynolds numbers for the flows in the pipelines are in the range from 4×10^3 to 1×10^5.

Derive an expression for the flow rate ratio Q_1/Q_2 assuming that minor losses can be neglected.

2.9 There is a pipeline with internal diameter $d = 10\,\text{mm}$. Compare the friction loss in the pipeline in the case A and B.

- Case A: Mineral hydraulic oil with a density of $\rho = 850\,\text{kg/m}^3$ and a viscosity of $\mu = 0.0272\,\text{Pa}\cdot\text{s}$ flows through the pipe at a flow rate of $Q = 25\,\text{L/min}$.
- Case B: High water-content fluid with a density of $\rho = 1000\,\text{kg/m}^3$ and a viscosity of $\mu = 0.0012\,\text{Pa}\cdot\text{s}$ flows through the pipe at a flow rate of $Q = 25\ \text{L/min}$.

2.10 Show graphically the relationship between the eccentricity rate $\bar{e} = e/h$ and leakage flow rate Q through the bore-spool clearance under the following conditions.

- Bore diameter: $2R = 20\,\text{mm}$
- Mean clearance between the spool and bore: $h = 25\,\mu\text{m}$
- Length of spool: $\ell = 15\,\text{mm}$
- Eccentricity: $e = 0 \sim h$
- Fluid viscosity: $\mu = 0.026\,\text{Pa}\cdot\text{s}$
- Pressure difference between the spool ends: $\Delta p = 20\,\text{MPa}$

2.11 A hydrostatic bearing has a disk diameter $2r_2 = 60\,\text{mm}$ and a bore diameter $2r_1 = 20\,\text{mm}$ (see Fig. 2.34), and it supports a load of 15 kN. A fixed displacement pump supplies fluid to the bearing at a flow rate of $Q = 6\,\text{L/min}$. The fluid has a viscosity $0.02\,\text{Pa}\cdot\text{s}$. Calculate the supplied pressure p_s and the clearance h in the hydrostatic bearing.

2.12 The disk with radius r_0 shown in Fig. 2.44, approaches the board at a velocity v, with the disk surface parallel with the board surface. Then

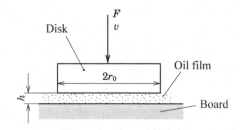

Fig. 2.44   Squeeze effect.

the oil in the clearance between the disk and board is pushed out, and the force due to the pressure in the clearance withstands the contact of both surfaces. The phenomenon is called the squeeze effect.

(1) Derive the pressure distribution equation $p(r)$ in the clearance by using Eq. (2.150).

(2) Derive the equation of the force $F(h)$ withstanding the contact with the lower surface.

Chapter 3

Hydraulic Components

Hydraulic systems comprise:

- pumps to transform the energy of prime movers into hydraulic energy,
- pipelines with various types of valves to control the hydraulic energy,
- actuators to transform hydraulic energy into work, and
- system components such as reservoirs, accumulators, filters, strainers, and heat exchangers.

This chapter provides a detailed overview of these hydraulic components.

3.1 Hydraulic Pumps and Motors

3.1.1 *Types of Hydraulic Pump and Motor*

A pump is a device that transforms the shaft power of a prime motor into fluid power. Generally, hydraulic pumps are positive displacement pumps. These have a pumping action that captures the working fluid in the pump chambers, thus increasing the pump chamber volume, and then transmits the pump shaft power to the trapped fluid. At the same time the pump discharges the trapped fluid by reducing the pump chamber volume. Positive displacement pumps are classified in terms of their structure as gear pumps, vane pumps and piston pumps. They are suitable for use at high pressures, since they discharge the exact specified amount of the fluid for each revolution of the driving shaft even if there are any load variations.

A hydraulic motor is a device that transforms fluid power into the rotational power of the motor shaft, though it has a geometry similar

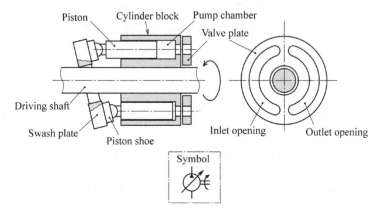

Fig. 3.1   Swash plate type axial piston pump.

to a positive displacement pump. That is, if pressurized working fluid is supplied to the hydraulic pump, then, in principle, it serves as a hydraulic motor transforming the fluid energy into the mechanical rotation energy.

Figure 3.1 illustrates a swash plate type axial piston pump. Several pistons are located in bores in a cylinder block, which is fixed to the pump shaft. At the end of each piston is a shoe that maintains a sliding contact with the swash plate as the block rotates. Thus, the end of each piston slides along a circular path on the swash plate. However, the swash plate is inclined at an angle to the cylinder block, and therefore the pistons reciprocate in the cylinder block bores. In the region where the bores of the cylinder block are located at the inlet port of the valve plate, the reciprocating pistons increase the pump chamber volumes, and the working fluid is sucked into the pump chambers through the inlet port. Similarly, the pump chamber volume decreases in the region where the cylinder block bores are transferred to the outlet port on the valve plate, and the fluid in the pump chamber is discharged in process of decreasing the pump chamber volume.

Axial piston pumps may be the swash plate type or the bent axis type. Figure 3.2 illustrates a bent axis type axial piston pump. Several pistons in the cylinder block are connected to a plate with a bent axis by means of a shaft coupling such as a universal joint. Since the bent axis rotates while keeping a constant inclination against the centerline of the cylinder block, the pistons reciprocate in the cylinder block bores, thus providing a pumping action in the same way as the swash plate type.

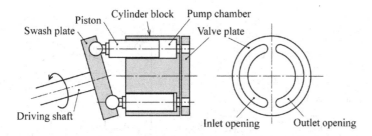

Fig. 3.2   Bent axis type axial piston pump.

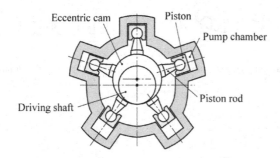

Fig. 3.3   Radial piston pump.

Figure 3.3 illustrates a radial piston pump with several piston-cylinders configured radially on an eccentric cylindrical cam. The rotation of the driving shaft reciprocates the pistons due to the cam movement so that the pump chamber volume varies. Working fluid is sucked into the pump chamber when the chamber volume increases, because the pressure reduction in the pump chamber opens the check valve in the suction line, as shown in Fig. 1.1. In contrast, the decrease in pump chamber volume makes the fluid discharge to the outlet line, because the increase of pressure in the pump chamber opens the check valve in the outlet line. The radial piston type of device is generally employed more in hydraulic motors than hydraulic pumps.

Figures 3.4(a) and (b) illustrate an external gear pump and an internal gear pump, respectively. In the external gear pump, the fluid at the pump inlet is captured in the pump chamber enclosed between the casing and gear space. This is because the inlet chamber volume increases in the process of the gear rotation. Then the captured fluid in the pump chamber is transferred to the outlet chamber and pushed out by reducing the outlet chamber volume. In the internal gear pump, the working fluid at the pump

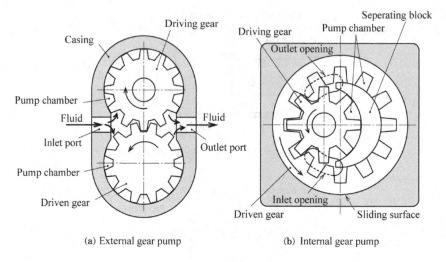

(a) External gear pump   (b) Internal gear pump

Fig. 3.4   Gear pumps.

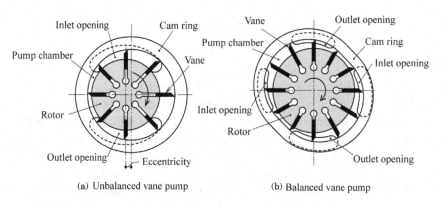

(a) Unbalanced vane pump   (b) Balanced vane pump

Fig. 3.5   Vane pumps.

inlet is captured into the pump chamber enclosed between the separating block and the gear space, and the pumping action is carried out in the same way as in an external gear pump.

Figures 3.5(a) and (b) illustrate an unbalanced vane pump and a balanced vane pump, respectively. Vanes are set into the radial slots of the rotor, so that they are able to slide in a radial direction. As the rotor rotates, the tip of each vane is pushed outwards against the inner surface of the cam ring by centrifugal force and by the hydraulic pressure in the spaces in the vane bottom.

In the unbalanced vane pump shown in Fig. 3.5(a), the center of the cam ring is located a small distance from the center of the rotor so that the pump chamber volume enclosed by adjacent vanes varies with the vane movement. The working fluid in the inlet chamber is brought into the pump chamber between adjacent vanes because the inlet chamber volume increases in the process of the vane movement. The trapped fluid is transferred to the outlet chamber by the movement of the vane, and as the outlet chamber volume reduces, so the fluid is discharged.

The unbalanced vane pump needs radial bearings to support the large load associated with the output pressure, because the inlet port is located opposite the output port. The unbalanced vane pump is a variable displacement pump with a simple geometry. It is possible to vary the output flow rate of such a pump by altering the eccentricity of the cam ring.

In the balanced vane pump shown in Fig. 3.5(b), outlet openings are symmetrically placed, so as to cancel the bearing load associated with the output high pressure. By using the cam ring with an inner shape similar to ellipse, the pumping action is achieved in a similar manner to that of an unbalanced vane pump.

The flow rate may be calculated using the theoretical fluid volume $V_p = 2\pi D_p$ that is swept in each revolution of the pump shaft. The volume V_p is called the displacement or swept volume and it represents the pump capacity. When the pump shaft rotates at the angular velocity ω, the theoretical pump flow rate Q is expressed as follows:

$$Q = D_p \omega, \qquad (3.1)$$

where $D_p = V_p/2\pi$ denotes the theoretical pump displacement volume per unit rotational angle of the pump shaft.

The rotational speed of the pump driven by an electric motor normally lies in the range from 1200 min^{-1} (revolutions per minute) to 1800 min^{-1}, and the normal rotational speed of the pump driven by a heat engine is in the range from 3500 to 10000 min^{-1}.

The pump transforms shaft power $T_a \omega$ into fluid power pQ_a. Thus, the overall efficiency η_p of a pump is defined by the ratio of the output fluid power pQ_a to the shaft power $T_a \omega$

$$\eta_p = \frac{pQ_a}{T_a \omega}, \qquad (3.2)$$

Table 3.1   Hydraulic pump performance.

Classification	Type	Swept volume [cm³/rev]	Maximum pressure [MPa]	Maximum rotational speed [min⁻¹]	Overall efficiency [%]
Gear pump	External gear	1–500	1–25	900–4000	75–85
	Internal gear	1–500	1–30	1200–4000	65–90
Vane pump	Balanced	1–350	3.5–40	1200–4000	70–90
	Unbalanced	10–230	3.5–21	1200–1800	60–70
Axial piston pump	Bent axis	10–1000	21–45	750–3600	88–95
	Swash plate	4–500	21–45	750–3600	85–90
Radial piston pump		6–500	14–25	1000–1800	85–90
Reciprocal piston pump		1–80	30–50	1000–1800	85–90

(a) Hydraulic pump

One directional flow
Electric motor drive
Fixed displacement
One directional rotation

(b) Hydraulic pump

One directional flow
Prime motor drive
Variable displacement
One directional rotation

(c) Hydraulic motor

Fixed displacement
Both directional rotations
Double axis

Hydraulic pump/motor

Variable displacement
Single axis
One directional rotation

Fig. 3.6   Standard graphic symbols of pumps and motors.

where Q_a denotes the actual effective output flow rate, which is obtained by subtracting the leakage flow from the theoretical flow rate Q as shown Eq. (3.8). In addition, p is the pump output pressure, T_a is the shaft torque to drive the pump and ω is the angular velocity of the pump shaft. Table 3.1 shows the typical characteristics of various types of hydraulic pumps, and Fig. 3.6 shows the standard graphic symbols of hydraulic pumps and motors.

Table 3.2 shows the swept volumes V_p derived theoretically from the geometrical volumes of the pump chambers. By designing a positive displacement pump so as to make the chamber volumes as small as possible and coincidentally make as many pump chambers as possible, the pressure fluctuation in the output flow can be reduced and its frequency increased. The number of the vanes or pistons in a pump is structurally restricted, so typically, the number of pistons is in the range from 5 to 9, and the number of vanes is around 13. It may be noted that when there are an odd number of vanes or pistons, the pressure fluctuations are smaller than when there are an even number of vanes or pistons.

Table 3.2 Swept volume of various type hydraulic pumps.

Classification	Theoretical swept volume V_p	Notation
Gear pump	$V_p \approx 2\pi b m^2 z$	b : Width of tooth m : Module z : Number of tooth
Unbalanced vane pump	$V_p \approx 2be(\pi D_c - zt)$	b : Width of vane z : Number of vane t : Thickness of vane e : Eccentricity between rotor and circular cam ring D_c : Diameter of circular cam ring
Balanced vane pump	$V_p \approx \pi b(D_l - D_s)\left(\dfrac{D_l - D_s}{2} - \dfrac{zt}{\pi}\right)$	D_l : Diameter of large arc on cam ring D_s : Diameter of small arc on cam ring
Swash plate type axial piston pump	$V_p = \dfrac{\pi d^2}{4} zD \tan \alpha$	d : Diameter of piston z : Number of piston D : Pitch diameter of piston on cylinder block α : Inclined angle e : Eccentricity between driving shaft and cam/crank
Bent axis type axial piston pump	$V_p = \dfrac{\pi d^2}{4} zD \sin \alpha$	
Radial piston pump (Reciprocal piston pump)	$V_p = \dfrac{\pi d^2}{2} ez$	

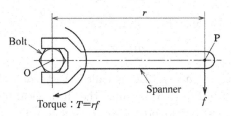

Torque : $T=rf$

Fig. 3.7   Definition of torque.

3.1.2 *Efficiency of Hydraulic Pumps and Motors*

3.1.2.1 *Hydraulic Pumps*

When a force f acts on the spanner with an arm length r as shown in Fig. 3.7, the torque T is defined as follows:

$$T = rf. \tag{3.3}$$

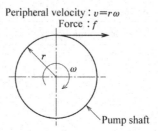

Fig. 3.8   Pump shaft.

Torque is defined as the tortional moment, and expressed in the same unit as energy, i.e. N.m though torque is independent of the rotational movement.

As explained in Section 1.1, the work per unit time is defined as power $W = fv$, where f is a force acting on an object with velocity v. As shown in Fig. 3.8, a shaft with radius r rotating at angular velocity ω has a peripheral velocity $v = r\omega$ and if a force f is applied at the shaft radius, then the power applied to the shaft is given by

$$W = fr\omega = T\omega. \tag{3.4}$$

Therefore, torque $T = fr$ and angular velocity ω in a rotational movement correspond to a force f and velocity v in a linear movement. In an ideal pump without energy losses, the pump shaft power $T\omega$ would equal the output fluid power pQ

$$T\omega = pQ. \tag{3.5}$$

Substituting Eq. (3.1) into Eq. (3.5) gives

$$T = D_p p. \tag{3.6}$$

The flow rate Q given in Eq. (3.1) is called the theoretical flow rate, and the torque T given in Eq. (3.6) is called the theoretical torque.

Figure 3.9 shows a fluid power source system. The actual torque T_a required to drive the pump shaft is denoted by adding the pump fluidic friction torque T_s to the theoretical torque T

$$T_a = T + T_s. \tag{3.7}$$

The actual effective output flow rate Q_a is obtained by subtracting the leakage flow rate Q_L from the theoretical flow rate Q

$$Q_a = Q - Q_L. \tag{3.8}$$

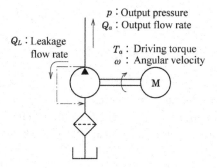

Fig. 3.9   Pump system.

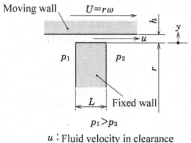

Fig. 3.10   Clearance model of pump.

As shown in Eq. (3.2), the overall efficiency of the pump is defined as the ratio of the effective fluid power to the power applied in the pump shaft. Therefore,

$$
\left.
\begin{aligned}
\eta_p &= \frac{Q_a}{Q}\frac{pQ}{T_a\omega} = \frac{Q_a}{Q}\frac{D_p p}{T_a} = \eta_{pv}\eta_{pT} \\[2mm]
\eta_{pv} &= \frac{Q_a}{Q} \\[2mm]
\eta_{pT} &= \frac{D_p p}{T_a} = \frac{T}{T_a}
\end{aligned}
\right\},
\qquad (3.9)
$$

where η_{pv} is called the volumetric efficiency of the pump, and η_{pT} is called the torque efficiency of the pump.

Let us suppose that the clearance dimensions in a pump are represented by a typical clearance model as shown in Fig. 3.10. Then, the fluidic frictional torque T_s is derived by applying Eq. (2.5)

$$T_s = bLr\mu \left[\frac{du}{dy}\right]_{y=0}, \tag{3.10}$$

where bL stands for the surface area of the typical clearance model with the gyration radius r and $U = r\omega$ is the typical velocity of the moving wall. Differentiating the fluid velocity u in Eq. (2.98) with respect to y at $y = 0$ gives

$$\left[\frac{du}{dy}\right]_{y=0} = \frac{U}{h} - \frac{1}{2\mu}\frac{dp}{dx}h. \tag{3.11}$$

Since the wall velocity is $U = r\omega$ and the pressure gradient dp/dx in the clearance flow is given as $dp/dx = -(p_1 - p_2)/L$ in Eq. (2.101), substituting Eq. (3.11) into Eq. (3.10) gives

$$T_s = \frac{r^2bL}{h}\mu\omega + \frac{rbh}{2}(p_1 - p_2). \tag{3.12}$$

Equation (3.12) denotes that the fluidic frictional torque T_s is given by terms representing sliding velocity $U = r\omega$ and differential pressure $p = p_1 - p_2$. The factors r^2bL/h and rbh in Eq. (3.12) involve clearance dimensions that are dependent on the individual pump geometry and the capacity. Since the swept volume $V_p = 2\pi D_p$ stands for the typical pump capacity, the first and second term in the right side of Eq. (3.12) can be denoted using $C_r D_p \mu\omega$ and $C_f D_p(p_1 - p_2)$, respectively, and dimensionless factors C_r and C_f are estimated experimentally for individual pumps

$$T_s = C_r D_p \mu\omega + C_f D_p p, \tag{3.13}$$

where $p = p_1 - p_2$ denotes the pressure difference between the inlet and the outlet of the pump. Recalling Eqs. (3.6), (3.7) and (3.13), the actual torque T_a is expressed as follows

$$T_a = D_p p + C_r D_p \mu\omega + C_f D_p p. \tag{3.14}$$

When the effect of the moving wall velocity on leakage can be neglected, the leakage flow rate Q_L through the clearance h with thickness width b is derived by applying Eq. (2.103)

$$Q_L = \frac{bh^3(p_1 - p_2)}{12\mu L}.$$ (3.15)

Since the factor $bh^3/(12L)$ in Eq. (3.15) depends on the individual pump geometry and capacity, it can be replaced by $C_s D_p$ and the dimensionless factor C_s is obtained experimentally for the individual pump. Then, Eq. (3.15) may be arranged as follows:

$$Q_L = C_s \frac{D_p p}{\mu}.$$ (3.16)

Recalling Eqs. (3.1), (3.8) and (3.16), an actual effective flow rate Q_a is defined as follows:

$$Q_a = D_p \omega - C_s \frac{D_p p}{\mu}.$$ (3.17)

Substituting Eq. (3.1) and (3.17) into Eq. (3.9) gives

$$\eta_{pv} = \frac{Q_a}{Q} = 1 - C_s \frac{p}{\mu \omega}.$$ (3.18)

In addition, substituting Eq. (3.1) and (3.14) into Eq. (3.9) gives

$$\eta_{pT} = \frac{T}{T_a} = \frac{1}{1 + C_f + C_r \frac{\mu \omega}{p}}.$$ (3.19)

Recalling Eqs. (3.18) and (3.19), an overall pump efficiency is defined as follows:

$$\eta_p = \eta_{pv} \eta_{pT} = \frac{1 - C_s \frac{p}{\mu \omega}}{1 + C_f + C_r \frac{\mu \omega}{p}}.$$ (3.20)

3.1.2.2 *Efficiency of Hydraulic Motors*

A hydraulic motor is a component that transforms fluid power into mechanical rotational power. Figure 3.11 illustrates the motor system that generates an effective torque T_{am} and rotates a load at angular velocity ω by supplying pressurized fluid into the hydraulic motor. The theoretical fluid volume $V_m = 2\pi D_m$ required to ensure a revolution of the motor is called the swept volume of the motor, and it is used as a factor representing the motor capacity.

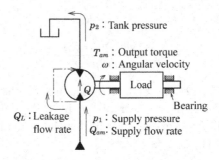

Fig. 3.11   Motor load system.

The overall efficiency of the motor η_m is defined by the ratio of the motor shaft power $T_{am}\omega$ to the supplied fluid power pQ_{am}

$$\eta_m = \frac{T_{am}\omega}{pQ_{am}} = \frac{Q}{Q_{am}}\frac{T_{am}\omega}{pQ} = \eta_{mv}\eta_{mT}, \qquad (3.21)$$

where $\eta_{mv} = Q/Q_{am}$ and $\eta_{mT} = T_{am}\omega/(pQ)$ are called the volumetric efficiency and the torque efficiency of the motor, respectively. Rotating the motor shaft at an angular velocity ω, the theoretical flow rate Q required to rotate the motor is described as follows:

$$Q = D_m\omega. \qquad (3.22)$$

The theoretical shaft power $T\omega$ equals the fluid power pQ in an ideal motor without any losses. Recalling $T\omega = pQ$ and Eq. (3.22), theoretical torque T is defined as follows:

$$T = D_mp. \qquad (3.23)$$

The actual effective torque of the motor T_{am} is given by subtracting the fluid friction torque T_s from the theoretical torque T

$$T_{am} = T - T_s. \qquad (3.24)$$

The flow rate required to drive the hydraulic motor Q_{am} is obtained by adding the leakage flow rate Q_L to the theoretical flow rate Q

$$Q_{am} = Q + Q_L. \qquad (3.25)$$

The fluid friction torque T_s is defined as $T_s = C_rD_m\mu\omega + C_fD_mp$ or as the fluid friction torque of the pump in Eq. (3.13). Substituting T_s and

Eq. (3.23) into Eq. (3.24) gives

$$T_{am} = D_m p - C_r D_m \mu \omega - C_f D_m p. \tag{3.26}$$

The leakage flow rate Q_L is defined by $Q_L = C_s(D_m p/\mu)$ or as the leakage flow rate of the pump in Eq. (3.16). Substituting $Q_L = C_s(D_m p/\mu)$ and Eq. (3.22) into Eq. (3.25) gives

$$Q_{am} = D_m \omega + C_s \frac{D_m p}{\mu}. \tag{3.27}$$

Recalling Eqs. (3.21), (3.26) and (3.27), the efficiencies of the hydraulic motor are derived accordingly:

$$\eta_{mv} = \frac{Q}{Q_{am}} = \frac{1}{1 + C_s \frac{p}{\mu \omega}}, \tag{3.28}$$

$$\eta_{mT} = \frac{T_{am}}{T} = 1 - C_f - C_r \frac{\mu \omega}{p}, \tag{3.29}$$

$$\eta_m = \eta_{mv} \eta_{mT} = \frac{1 - C_f - C_r \frac{\mu \omega}{p}}{1 + C_s \frac{p}{\mu \omega}}. \tag{3.30}$$

As shown in Eqs. (3.28) to (3.30), it is clear that hydraulic motor efficiencies are the functions of the variable $\mu \omega/p$.

[**Example 3.1**]

As shown in Fig. 3.12, the system pressure is $10\,\text{MPa}$ and the motor shaft power is $W = 7.5\,\text{kW}$ when the motor shaft rotates under load at angular velocity $\omega = 120\,\text{rad/s}$. The flow rate $Q_L = 3\,\text{L/min}$ is discharged through the relief valve into the reservoir. The pressure loss through the directional control valve is 0.2 MPa and the pipe line pressure loss is 0.2 MPa. Calculate the swept volume $2\pi D_m$ of the hydraulic motor, the pump effective flow rate Q_p, and the pump shaft power W_i assuming the efficiencies of the pump and the motor as follows:

The torque efficiency of the pump: $\eta_{pT} = 0.9$;
The volumetric efficiency of the pump: $\eta_{pv} = 0.96$;
The torque efficiency of the motor: $\eta_{mT} = 0.84$;
The volumetric efficiency of the motor: $\eta_{mv} = 0.94$.

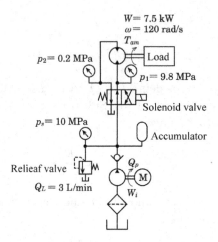

Fig. 3.12   Hydraulic system.

[Solution 3.1]

The effective torque T_{am} is obtained from Eq. (3.4)

$$T_{am} = \frac{W}{\omega} = \frac{7.5 \times 10^3}{120} = 62.5 \text{ N} \cdot \text{m}.$$

Equations (3.29) and (3.23) yield the motor theoretical torque T and the motor swept volume D_m as follows:

$$T = \frac{T_{am}}{\eta_{mT}} = \frac{62.5}{0.84} = 74.4 \text{ N} \cdot \text{m},$$

$$D_m = \frac{T}{p} = \frac{74.4}{(10 - 0.4) \times 10^6} = 7.75 \times 10^{-6} \text{m}^3/\text{rad} = 7.75 \text{cm}^3/\text{rad}.$$

Therefore, the swept volume per unit revolution is

$$2\pi D_m = 2 \times 3.14 \times 7.75 = 48.7 \text{ cm}^3/\text{rev}.$$

Recalling Eq. (3.22), the motor theoretical flow rate Q becomes

$$Q = D_m \omega = (7.75 \times 10^{-6}) \times 120 = 0.93 \times 10^{-3} \text{m}^3/\text{s}.$$

Therefore, the pump effective flow rate Q_p is given as

$$Q_p = \frac{Q}{\eta_{mv}} + Q_L = \frac{0.93 \times 10^{-3}}{0.94} + \frac{3 \times 10^{-3}}{60} = 1.04 \times 10^{-3} \text{m}^3/\text{s}.$$

The pump output power W_p and the pump shaft power W_i are required as follows:

$$W_p = Q_p p_s = (1.04 \times 10^{-3}) \times (1 \times 10^7) = 10.4\,\text{kW}.$$

$$W_i = \frac{W_p}{\eta_{pT}\eta_{pv}} = \frac{10.4}{0.9 \times 0.96} = 12.0\,\text{kW}.$$

3.1.3 *Pump Characteristics*

Pump characteristics such as the flow rate characteristics, the output power, the shaft power, the shaft torque and the efficiencies for typical operating conditions are generally provided in the manufacturer's specifications and product catalogues. Figure 3.13 shows the performance of the fixed displacement pump typically used in an aircraft. It is measured under the operating conditions whereby the rotational speed is 3500 min^{-1}, the pressure at the suction port is 0.25 MPa and the pump is driven using MIL-H5606C hydraulic fluid at a temperature of 70°C. Estimating the factors C_s, C_f and C_r based on the pump characteristics and assuming the variable driving conditions, the pump efficiencies η_p, η_{pv} and η_{pT} are functions of the variable $\mu\omega/p$. The value of the factor C_s is estimated by substituting values of the parameter $p/(\mu\omega)$ for the specific condition into Eq. (3.18). The values of the factors C_f and C_r are estimated by solving the simultaneous equations that are obtained by substituting the values of $\mu\omega/p$ relevant to the two specified conditions into Eq. (3.19). For instance, using the pump efficiencies at the pressures $p_o = 21$ MPa and $p_o = 3.5$ MPa moderately selected on the basis of Fig. 3.13, the values of C_s, C_f and C_r can be estimated as follows.

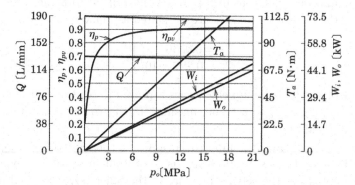

Fig. 3.13 Characteristics of a fixed displacement pump.

Table 3.3 Parameter values.

p_o [MPa]		21	3.5
η_p		0.910	0.825
η_{pv}		0.965	0.995
η_{pT}		0.943	0.830
$\dfrac{\mu\omega}{p} = \dfrac{\mu\omega}{p_o - p_i}$		1.17×10^{-7}	7.35×10^{-7}

The overall pump efficiency η_p, the volumetric efficiency η_{pv} and the torque efficiency η_{pT} corresponding to the output pressure $p_o = 21$ MPa and $p_o = 3.5$ MPa are derived from Fig. 3.13 and Eq. (3.9). Table 3.3 summarizes the efficiencies and the parameter values of $\mu\omega/p = \mu\omega/(p_o - p_i)$ calculated using $\mu = 0.00668$ Pa·s, the angular velocity $\omega = 366$ rad/s and the inlet pressure $p_i = 0.25$ MPa. Substituting the values of η_{pv} and $\mu\omega/(p_o - p_i)$ corresponding to $p_o = 21$ MPa or $p_o = 3.5$ MPa into Eq. (3.18) gives

$$C_s = 0.413 \times 10^{-8}. \tag{3.31}$$

Substituting the values of η_{pv} and $\mu\omega/(p_o - p_i)$ corresponding to $p_o = 21$ MPa into Eq. (3.19) gives

$$0.943 C_f + 0.111 \times 10^{-6} C_r = 0.057. \tag{3.32}$$

Substituting the values of η_{pv} and $\mu\omega/(p_o - p_i)$ corresponding to $p_o = 3.5$ MPa into Eq. (3.19) gives

$$0.83 C_f + 0.625 \times 10^{-6} C_r = 0.17. \tag{3.33}$$

The simultaneous equations of Eq. (3.32) and Eq. (3.33) being solved, the values of the factors C_f and C_r are estimated as follows:

$$\left. \begin{array}{l} C_f = 0.0336 \\ C_r = 0.227 \times 10^6 \end{array} \right\}. \tag{3.34}$$

Figure 3.14 shows the pump efficiencies arranged in relation to the parameter $\mu\omega/p = \mu\omega/(p_o - p_i)$ and recalling the factor values obtained in Eqs. (3.31) and (3.34).

A fixed displacement pump delivers almost a constant flow rate regardless of the output pressure. However, as the output pressure in a hydraulic system approaches the setting pressure of the relief valve installed in the system, a part of the pump displacement begins to pass through the relief

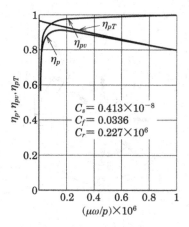

Fig. 3.14   Pump efficiencies.

valve, and then almost all the pump displacement is exhausted to the reservoir through the relief valve. The energy of the fluid passing through the throttle of the relief valve is transformed into heat energy without accomplishing effective work.

However, the relief valve has an important role to play in regulating the maximum system pressure. Therefore, in the context of energy savings it is preferable that the swept volume should be controlled so as to become nearly zero in accordance with the setting pressure of the relief valve. In order to save energy, pressure compensated variable displacement pumps are widely used in hydraulic systems even though they can be present in electro-hydraulic systems in which the piston pump displacement is controlled by varying the inclination of the swash plate according to the pressure signal.

Figure 3.15 illustrates a pressure compensated variable displacement vane pump. The cam ring in the pump is supported by a thrust block, a screw stopper and a governor piston. Rotating the pump shaft, a thrust F occurs as shown in Fig. 3.15 because the output pressure acts on the inner surface of the cam ring. Though the thrust F is quite large even in a small vane pump, the y-direction thrust component F_y comprises a large portion of the thrust, and it is supported by the thrust block. The screw stopper limits the travel of the cam ring in the output pressure region, where the governor spring force overcomes the x-direction thrust component F_x. Thus, the pump delivers nearly a constant flow rate. However, the output flow rate does decrease when the eccentricity between the rotor and cam

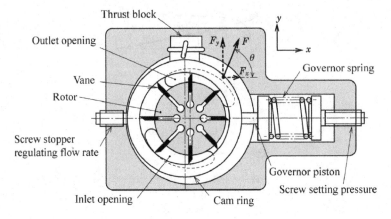

Fig. 3.15 Pressure compensated variable displacement vane pump.

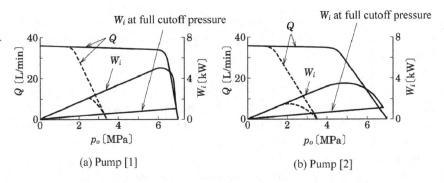

(a) Pump [1]

(b) Pump [2]

Fig. 3.16 Characteristics of pressure compensated variable displacement vane pump.

ring decreases. This happens when the output pressure grows so high as to overcome the thrust F_x of the initial governor spring force. As the output pressure continues to rise, the output flow rate eventually becomes nearly zero at the full cutoff (dead head). The full cutoff stands for the condition when the pump delivers only the flow rate corresponding to the leakage flow rate and then the thrust F_x balances the governor spring force dependent on the maximum deflection.

Figure 3.16 shows the flow rate characteristics of two pressure compensated variable displacement vane pumps which are conformed for different applications. The output flow rate of the pump [1] decreases more rapidly in the output pressure region where the thrust force F_x exceeds the governor spring force set by the initial deflection.

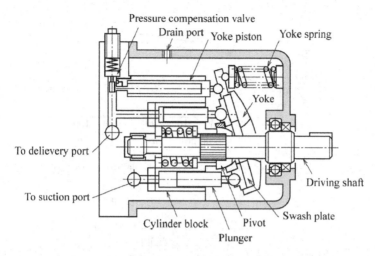

Fig. 3.17   Pressure compensated variable displacement piston pump.

Figure 3.17 illustrates a pressure compensated swash plate type axial piston pump. If the output pressure approaches the setting pressure, then the pressure compensator valve begins to slightly open the port opening to the yoke cylinder chamber. Then the yoke piston operates so as to decrease the inclination of the swash plate. This reduces the plunger stroke, and the pump displacement decreases rapidly as the output pressure increases further. Presently the full cutoff occurs at the setting pressure.

Piston type and vane type pumps generate pressure fluctuations. These are caused by the excess pre-compression of the working fluid confined in the pump chamber in the transient process of fluid transfer from inlet region to outlet region. Also, gear pumps generate pressure fluctuations caused by the excess compression of the confined fluid in the overlapping teeth ditches. Such pressure fluctuations tend to excite the pump structure, producing unwanted noise. Extensive research has been undertaken to decrease pump noise, and a noise reduction of over 10 dB has been achieved in the last decades. Generally, a pump with noise characteristics below 60 dB is regarded as a quiet pump. Figure 3.18 shows the noise levels produced by variable displacement piston pumps with different capacities, based on data supplied by the manufacturers.

Pumps with capacities from 1000 L/min to 0.3 L/min can be used in the maximum pressure range from 40 MPa to 7 MPa in several pump configurations. The pressure range can be increased by using pumps in stages. A multistage pump is engineered by connecting the first stage pump

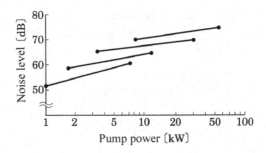

Fig. 3.18   Noise characteristics of variable displacement piston pumps.

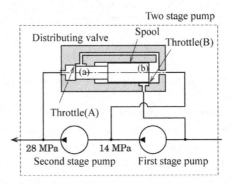

Fig. 3.19   Pressure sharing in staged pump.

outlet to the inlet of the second stage pump driven by the same shaft. In a two-stage pump configured in this way, the maximum output pressure becomes twice as high as that in a single stage.

Figure 3.19 illustrates the valve dividing the pressure between the first stage and the second stage in a multistage pump. The spool area (b) facing the first stage outlet is designed so as to be twice as large as the spool area (a) facing the second stage outlet. Therefore, when the second stage outlet pressure exceeds double the first stage outlet pressure, the spool moves right, and a part of the second stage displacement is bypassed through the throttle (A) into the first stage outlet. In contrast, when the first stage outlet pressure exceeds half of the second stage outlet pressure, the spool moves left, and a part of the first stage displacement is bypassed through the throttle (B) into the first stage inlet.

In a multiple-pump comprising multiple parallel connection pumps driven by the same shaft, the outlet ports of the pump may be connected to the same hydraulic circuit or to different ones. Figure 3.20 illustrates various types of multiple connection pumps using graphic symbols.

(a) Multiple-pump used as two independent pumps with separated suction ports

(b) Multiple-pump used as two independent pumps with common suction port

(c) Multiple-pump used as one pump

Fig. 3.20 Multiple-pumps.

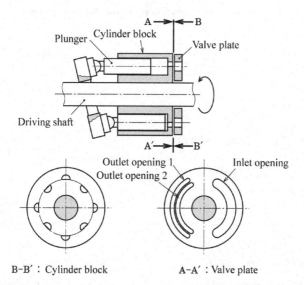

B–B′ : Cylinder block   A–A′ : Valve plate

Fig. 3.21 Split swash plate type axial piston pump.

Figure 3.21 illustrates a split flow swash plate type axial piston pump. The cylinder block has two series of semicircular block end bores, and the two series of bores are arranged in such a way as to open the valve plate openings 1 and 2 during the rotation of the cylinder block. Therefore, the pump has the two independent power sources with the same displacement. The split flow type pump is used in the power sources for synchronized operations such as power shovels in construction machines, because it is light and compact.

3.2 Hydraulic Actuators

Hydraulic components that transform fluid power into mechanical power are called hydraulic actuators. They are classified as shown in Table 3.4.

Table 3.4   Classification of hydraulic actuators and standard graphic symbols.

	Type of actuator		Symbol
Hydraulic motor	Fixed displacement	One directional flow	
		Bi-directional flow	
	Variable displacement	One directional flow	
		Bi-directional flow	
Rotary motor			
Hydraulic cylinder	Single acting		
	Double acting	Single rod	
		Double rod	

3.2.1 *Hydraulic Motors*

Hydraulic motors are rotational actuators with a geometry similar to that
of positive displacement pumps. Motors are provided with a case drain to
protect the shaft seals because the outlet pressure of the motor is higher
than atmospheric pressure. Therefore, the types of hydraulic motors correspond to the types of pumps, and there are also hydraulic pump-motors
designed to be convertible from motors to pumps. Axial piston motors are
widely advertised as combination pump-motors.

Let us consider the principle of hydraulic motors assuming that the
downstream flow is fed to a tank under atmospheric pressure. Working fluid is supplied to the gear motor at the supply pressure p_s as
shown in Fig. 3.22. The pressure acting on the teeth surfaces at the locations a_1 and a_2 generates the clockwise torque to rotate the motor shaft,
whereas the pressure acting on the interlocked teeth surfaces at locations
b_1 and b_2 generates a counterclockwise torque. However, the motor shaft
rotates clockwise because the former torque is about twice as large as the

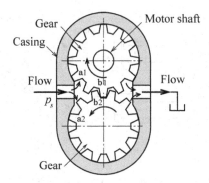

Fig. 3.22　Gear motor.

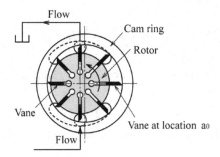

Fig. 3.23　Vane motor.

latter due to the difference between the surface areas receiving the supply pressure.

Figure 3.23 illustrates a vane motor in which the supply pressure acting on the vane at the location a_0 produces the counterclockwise torque rotating the motor shaft.

Figure 3.24 illustrates the operating principle of a swash plate type axial piston motor, where θ denotes the angle presenting the location of a plunger from Top Dead Center, R is the distance between the center of the plunger and center of the motor shaft, α is the inclined angle of the swash plate, and A is the plunger area receiving effectively the supply pressure p_s.

Let us consider the plunger force acting on the swash plate at an angle θ. The plunger force $f = Ap_s$ involves a force component $f_1 = f \sin \alpha$ along with the swash plate surface and a force component $f_2 = f \cos \alpha$ normal to it. The torque ΔT_m rotating the motor shaft is described as $fR \tan \alpha \cos(\theta - \pi/2)$ because the arm length is $R \cos(\theta - \pi/2)$. Therefore, the total plunger

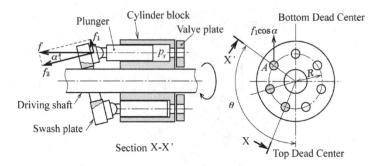

Fig. 3.24   Swash plate type axial piston motor.

torque T_m is expressed as follows:

$$T_m = p_s A R \tan\alpha \sum_{k=1}^{z_0} \sin\left[\theta + \frac{2\pi(k-1)}{z}\right].$$ (3.35)

When the number of plungers z is even, z_0 is given as follows:

$$z_0 = \frac{z}{2}$$ (3.36)

and when the number of the plunger z is odd, z_0 is given as

$$\left.\begin{array}{l} 0 \le \theta < \dfrac{\pi}{z} : z_0 = \dfrac{z+1}{2} \\[2mm] \dfrac{\pi}{z} \le \theta < \dfrac{2\pi}{z} : z_0 = \dfrac{z-1}{2} \end{array}\right\}.$$ (3.37)

Typical industrial axial piston motors are driven in the speed range from 900 to 4000 $\min^{-1}$. Recent developments in technology make the lower limit motor speed of 100 $\min^{-1}$ feasible for middle-speed/middle-capacity gear motors, and make the upper limit speed of 10000 $\min^{-1}$ feasible for high speed small capacity gear motors.

Radial piston motors are used as low-speed/high-torque motors in cranes, rolling mills, excavators, etc. Figure 3.25 illustrates the radial piston motor with a pentagon cam serving as a crank function to rotate the motor shaft. The crank arm length is the distance e from the pentagon center to the motor shaft center. The motor shaft has a sleeve dividing the cylindrical space in the pentagon-cam into two chambers. The chamber facing the a-side sleeve surface is trafficked to the high-pressure port, whereas the other chamber facing the b-side sleeve surface is trafficked to the exhaust port.

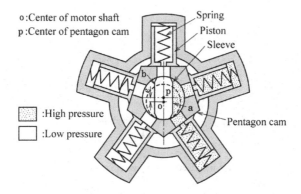

Fig. 3.25   Radial piston motor.

The motor shaft rotates in the pentagon cam around the center o due to the crank arm torque produced by the radial piston-cylinders into which the supplied fluid flows through the a-side sleeve chamber. Then the pentagon cam moves around the shaft center o without changing the attitude of the pentagon cam so that the radial pistons facing the b-side sleeve chamber discharge the fluid in the chambers.

Figure 3.26 shows the piston forces corresponding to rotational angle θ and acting when the motor shaft rotates counterclockwise. As shown in Fig. 3.26(a), the fluid flows into the piston chambers through port 1, port 2 and port 3 under the rotational angle range from $0°$ to $36°$, and these fluids transmit the piston forces $F = pA$ to the pistons with the area A. The piston force component normal to the cam crank arm in port 1, port 2 and port 3 becomes $pA\sin\theta$, $pA\cos(\theta - 18°)$ and $pA\sin(36° - \theta)$, respectively. Therefore the total piston torque T_m is defined as follows, since the action lines of the piston forces F pass through the pentagon center point p

$$0° \leq \theta \leq 36° : \ T_m = pAe[\sin\theta + \cos(\theta - 18°) + \sin(36° - \theta)]. \quad (3.38)$$

For rotational angles ranging from $36°$ to $72°$ as shown in Fig. 3.26(b), the ports 1 and 2 are kept in the high pressure end whilst port 3 is transferred to the side opening of the exhaust port. Then the total piston torque T_m is given as follows:

$$36° \leq \theta \leq 72° : \ T_m = pAe[\sin\theta + \cos(\theta - 18°)]. \quad (3.39)$$

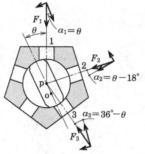

(a) $0 \leq \theta \leq 36°$: High pressure ports 1, 2 and 3

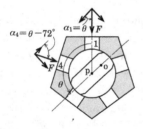

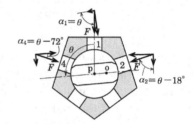

(b) $36° \leq \theta \leq 72°$: High pressure ports 1 and 2   (c) $72° \leq \theta \leq 108°$: High pressure ports 1, 2 and 4

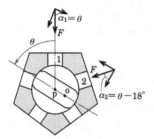

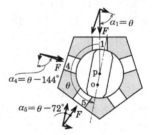

(d) $108° \leq \theta \leq 144°$: High pressure ports 1 and 4   (e) $144° \leq \theta \leq 180°$: High pressure ports 1, 4 and 5

Fig. 3.26   Radial piston forces for rotation angle.

The total torque T_m in each rotation angle region is derived by considering the piston forces as shown in Fig. 3.26(c)–(e)

$$
\left.
\begin{aligned}
72° \leq \theta \leq 108° : \ & T_m = pAe[\sin\theta + \cos(\theta - 18°) + \sin(\theta - 72°)] \\
108° \leq \theta \leq 144° : \ & T_m = pAe[\sin\theta + \sin(\theta - 72°)] \\
144° \leq \theta \leq 180° : \ & T_m = pAe[\sin\theta + \sin(\theta - 72°) + \sin(\theta - 144°)]
\end{aligned}
\right\}.
$$

$$(3.40)$$

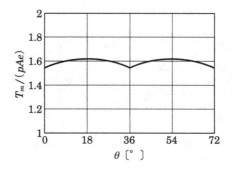

Fig. 3.27   Dimensionless torque of radial piston motor with pentagon cam.

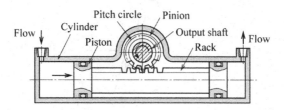

Fig. 3.28   Piston-rack and pinion type rotary motor.

Figure 3.27 shows the motor torque characteristics derived from Eqs. (3.38) and (3.39). The dimensionless torque $T_m/(pAe)$ changes in the 36° cycle of the rotational angle and the torque fluctuations are kept within 6.5%.

3.2.2 *Rotary Motors*

Rotary motors are rotational actuators whose working angles are usually limited within 360°. They are applied extensively in industry in clamping, positioning and turning devices. Piston type rotary motors transform linear power into rotary power by means of devices such as the rack and pinion mechanism, the helical spline and the chain-sprocket wheel.

Figure 3.28 illustrates a piston type rotary motor that transmits the limited rotary movements to the pinion shaft by actuating the piston with a rack on either side. The torque T_m is defined as follows:

$$T_m = \eta A p r_p, \tag{3.41}$$

where A, p, r_p and η are the piston area, the differential pressure acting on the piston, radius of the pinion pitch circle and the overall efficiency of the rotary motor, respectively.

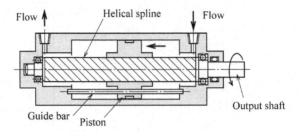

Fig. 3.29   Piston-helical spline type rotary motor.

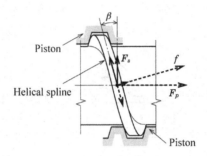

Fig. 3.30   Forces acting on helical spline.

Figure 3.29 illustrates a piston type rotary motor using a helical spline mechanism. The piston is supported by guide bars so that it should not rotate with the spline shaft. The helical spline type rotary motor is used as a low-torque/low-speed motor, since the inner leakage through the screw gap becomes too large as the supply pressure increases.

Figure 3.30 illustrates the forces acting on the spline, where the piston thrust force F_p gives a steady state rotation to the spline shaft with the load force F_s. Then the reaction force f acting on the sliding surface of the spline is expressed as follows:

$$f = F_s \sin \beta + F_p \cos \beta, \qquad (3.42)$$

where β denotes the helix angle.

The resultant sum of the force components along the helical spline surface should be zero because the force balance is maintained during the steady state rotation of the spline shaft. Accordingly, Eq. (3.43) is derived

$$F_p \sin \beta - F_s \cos \beta - \mu f = 0, \qquad (3.43)$$

where μ is the friction coefficient. Utilizing the thrust efficiency η_f responsible for the friction loss, the piston force F_p is expressed as follows:

$$F_p = \eta_f A p, \tag{3.44}$$

where p and A are the differential pressure and the piston area, respectively. The output torque T_m is obtained recalling Eqs. (3.42)–(3.44):

$$T_m = F_s r_p = \eta_f \frac{pAd_p\left(\sin\beta - \mu\cos\beta\right)}{2\left(\cos\beta + \mu\sin\beta\right)}, \tag{3.45}$$

where $d_p = 2r_p$ denotes the pitch circle diameter of the spline.

Figure 3.31 illustrates a single-vane type rotary motor with two chambers divided by the vane. As the working fluid is supplied into the chamber A, the shaft rotates with the vane clockwise and the fluid in the chamber B is discharged. The seal between the vane and the cylinder chamber surface must be designed carefully because motor performance is diminished by leakage and friction loss.

Figure 3.32 shows a double-vane type rotary motor with four chambers divided by double-vanes. Chambers A_1 and B_1 open to chambers A_2 and B_2, respectively through the pilot passages. When working fluid is supplied into chambers A_1 and A_2, the motor shaft rotates clockwise and the working fluid in chambers B_1 and B_2 is discharged. The double-vane rotary motor efficiency η_m lies in the range from 0.90 to 0.95, and it is higher than in the single-vane rotary motor because the bearing load decreases due to the pressure balance around the double vane rotary shaft.

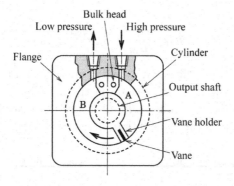

Fig. 3.31   Single-vane type rotary motor.

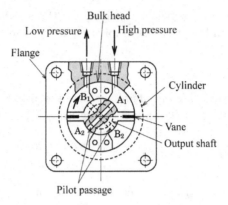

Fig. 3.32   Double vane type rotary motor.

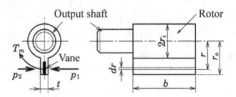

Fig. 3.33   Vane dimensions.

In the rotary motor with vane, shown in Fig. 3.33, the output torque T_m and the rotational speed ω are defined as follows:

$$T_m = \eta_{mT} n \int_{r_i}^{r_o} (p_1 - p_2) r b \, dr = \frac{n}{2} \eta_{mT} b (p_1 - p_2)(r_o^2 - r_i^2), \quad (3.46)$$

$$\omega = \eta_{mv} \frac{Q}{D_m} = \frac{2 \eta_{mv} Q}{n b (r_o^2 - r_i^2)}, \quad (3.47)$$

where η_{mt} is the torque efficiency, r_o and r_i are the outer and inner radiuses of the rotor, n is the number of vanes in the multi vane type motor, D_m is the swept volume per unit angle, Q is the flow rate flowing into the motor, and $p_1 - p_2$ is the differential pressure between the inlet port and the outlet port.

Figure 3.34 shows a double vane type rotary motor developed as the actuator to steer the nose wheel of a business jet aircraft. The stator is joined to the landing gear of the aircraft and the rotor drives the nose wheel of the aircraft.

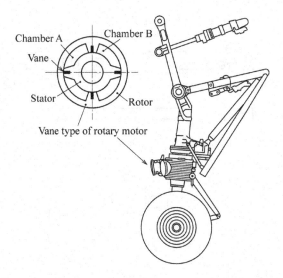

Fig. 3.34 Rotary motor used in steering wheel of aircraft.

3.2.3 *Hydraulic Cylinders*

3.2.3.1 *Introduction*

Hydraulic cylinders are actuators that transform fluid power into mechanical linear transfer power, and they are typically categorized either as single acting cylinders or double acting cylinders. Double acting cylinders have two chambers to supply or to discharge the working fluid, and make the piston move on each side. This is done by exchanging the supply port for the exhaust port alternately (switching between the supply and exhaust ports). In contrast, single acting cylinders have only one cylinder chamber to supply or discharge the working fluid, and the piston returns to its initial positions under the action of gravity or a spring. Rams belong to the single acting type cylinders, and the piston serves both as the piston and the rod. These are typically used for vehicle elevator devices, hydraulic press or dump cylinders, etc. The cylinders can be also categorized as the single rod end type and the double rod end type. Since the double acting single rod end type cylinders have only a rod through the one side cylinder chamber, they produce a greater force in extending the rod out of the cylinder chamber than in dragging the rod into the chamber. In contrast, double rod end type cylinders have the rod extending through each cylinder chamber, and produce the same force in both directions.

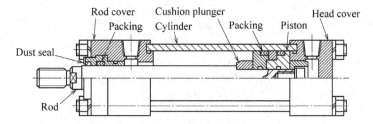

Fig. 3.35   Double acting single rod end type hydraulic cylinder.

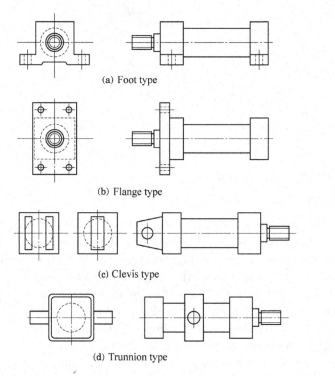

(a) Foot type

(b) Flange type

(c) Clevis type

(d) Trunnion type

Fig. 3.36   Types of cylinder mounting.

Figure 3.35 illustrates a double acting single rod end type cylinder that consists of the piston, the cylinder tube, the rod, the rod cover, the head cover, the packing seals and the cushion device. Figure 3.36 illustrates different types of mountings for the cylinders. They may be classified as: the foot type rigidly mounted with cylinder foot, the flange type rigidly mounted with the flange surface normal to the rod, and the trunnion type mounted so as to be moveable around the axis.

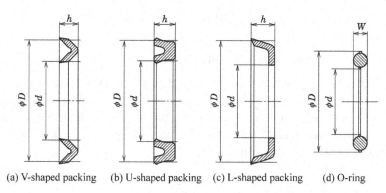

(a) V-shaped packing   (b) U-shaped packing   (c) L-shaped packing   (d) O-ring

Fig. 3.37   Cross sections of packing.

Figure 3.37 shows the cross-sections of the various types of packing to prevent internal and external leakages in the hydraulic components. The types of packing having the V-shape, U-shape and L-shape cross-sections shown in Fig. 3.37(a)–(c) are called lip type packing, and they are used as seals in dynamic parts such as pistons and rods in the cylinders. The inner diameter of the packing lip is slightly smaller than the rod diameter and the outer diameter of the packing lip is slightly larger than the inner diameter of the sealed cylinder tube, and moreover the packing lip contacts with sealed metal surface more tightly due to the fluid pressure acting on it.

Figure 3.37(d) shows the seal with an O-shape cross-section, called an O-ring. The O-ring is a typical compression type seal made of a nitrile compound rubber, and it is the most popular seal used in sealing dynamic surfaces and also static surfaces. The O-ring is set into a groove with a slightly smaller depth and slightly wider width than the O-ring cross-section diameter, and so the O-ring surface is compressed by the sealed surface and thus prevents leakage of the fluid. The sizes of the O-rings and the grooves are standardized. The types of the seals should be selected by considering the piston speed, the oil temperature, the pressure durability, the leakage, etc.

In general, the hydraulic cylinder is driven within the piston speed range from 8 mm/s to 400 mm/s. However, piston speeds lower than 20 mm/s often give rise to a phenomenon known as stick slip, in which the intermittent movement of the piston occurs due to coulomb friction.

In order to prevent stick slip in low speed cylinders, the rod surface and the inner surface of the cylinder tube are often plated with chromium, and special seals with low friction such as metal piston-rings or O-rings covered with the Teflon cap are often used.

The thrust efficiency η_c is defined as follows:

$$\eta_c = \frac{F}{A_1 p_1 - A_2 p_2}, \tag{3.48}$$

where F is the cylinder thrust force, A and p are respectively the piston area and the chamber pressure. Suffix numbers 1 and 2 denote the supply port side and the exhaust port side, respectively. Thrust efficiency η_c is generally within the region from 0.8 to 0.95. On the one hand, the volumetric efficiency η_v is defined as follows:

$$\eta_v = \frac{A_1 v}{Q}, \tag{3.49}$$

where v and Q are the piston speed and the supplied flow rate, respectively. However, the volumetric efficiency of the cylinder may be regarded as $\eta_v \approx 1$ because the leakage loss in a cylinder is negligibly small.

3.2.3.2 *Standard Cylinders*

There are Japanese industrial standards JIS B 8354 (Abolishion) that standardize the double acting hydraulic cylinders with regard to the inner geometry, the mounting method, the operating characteristics, the dimensions, etc.

In accordance with the industrial standards such as JIS B 8354, hydraulic cylinders are called standard cylinders. Standard cylinders are widely used for industrial machines since they offer such features as part interchangeability, fail-proof performances, cost-effectiveness and easy availability. The industrial standards are often referred to even in the design of the nonstandard cylinders. For instance, the inner diameters of nonstandard cylinders and rods are often designed in accordance with the industrial standards because of easy availability of seal components. Nonstandard hydraulic cylinders are widely used in the vehicle and aircraft engineering to meet the demands for the lighter and smaller elements. Figure 3.38 shows a nonstandard hydraulic cylinder using an aluminum alloy and the O-rings standardized for aircrafts, etc. The extracts from the JIS B 8354 having relevance to the hydraulic cylinder-tube and rod are shown in Tables 3.5–3.7.

3.2.3.3 *Cushion Devices*

Most cylinders have a cushion device within the head cover or the rod cover to prevent the piston hitting it at speed. Figure 3.39 illustrates a cushion device set in the head cover. When the cushion plunger has penetrated into

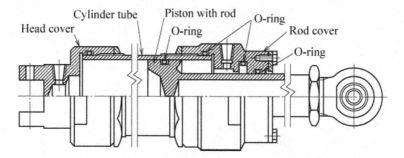

Fig. 3.38 Nonstandard cylinder used for aircraft.

Table 3.5 Standard cylinder-tube inner diameters and rod diameters [mm].

| Cylinder inner diameter | Symbols of rod diameter | | |
	A	B	C
32	22	18	14
40	28	22	18
50	36	28	22
63	45	36	28
80	56	45	36
100	70	56	45
125	90	70	56
140	100	80	63
160	110	90	70
180	125	100	80
200	140	110	90
220	160	125	100
250	180	140	110

Table 3.6 Allowable maximum pressure [MPa].

| Nominal pressure | Allowable pressure at head side | Allowable pressure at rod side | | |
| | | Symbol of rod diameter | | |
		A	B	C
7	9	15	13.5	11
14	18	18	18	14
21	27	25	25	—

Table 3.7   Piston speeds.

Cylinder diameter [mm]	Piston speed [m/s]
32–63	8–400
80–125	8–300
140–250	8–200

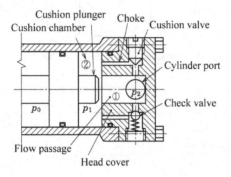

Fig. 3.39   Cushion device.

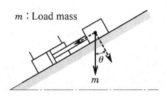

Fig. 3.40   Load condition.

the head cover bore, the plunger closes the flow path so that the fluid in the cushion chamber is forced to pass through the cushion valve. Then, the kinetic energy of the piston is transformed into fluid heat energy by the viscous friction of the fluid passing through the cushion valve throttle.

The loading situation of a mass connected to the rod end affects the allowable maximum piston speed at the time when the cushion device has begun to operate. Therefore, the limited piston speeds V_L are regulated in JIS B 8354 by the equivalent mass in accordance with the loading condition of the mass defined by Fig. 3.40 and in the following Eq. (3.50). Figure 3.41 shows the limited piston speed V_L in relation to the equivalent mass

$$m_e = n \left[\frac{A_1 p_0 - f}{g} + m \left(1 \pm \sin \theta \right) \right], \qquad (3.50)$$

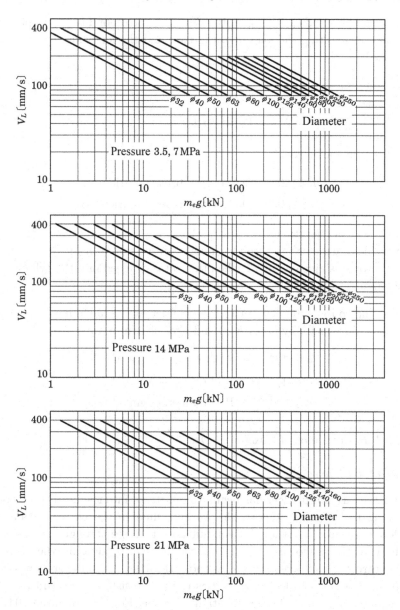

Fig. 3.41 Limited piston speeds.

where the plus and the minus in the sign "$\pm$" correspond to the respective lifting and lowering the load onto the base and the following designations are used:

m : Load mass
A_1 : Piston area
f : Load friction force ($f = \mu m g \cos\theta$)
p_0 : Supply pressure
μ : Friction coefficient

$$\mu \approx 0.03 \quad \text{(Boundary lubrication friction)}$$
$$\mu \approx 0.002\text{--}0.006 \quad \text{(Roller-bearing friction)}$$

g : Gravity acceleration
θ : Inclined angle of the base to support load as shown in Fig. 3.40
n : Factor n is given as $n = 1$ when the rod is dragged into the cylinder in Fig. 3.40.

When the rod is extended out of the cylinder in Fig. 3.40, it is expressed by the rod diameter symbols A, B and C as shown in Table 3.6.

$$n = 1/2 \quad \text{(Rod diameter symbol A)};$$
$$n = 1/1.45 \quad \text{(Rod diameter symbol B)};$$
$$n = 1/1.25 \quad \text{(Rod diameter symbol C)}.$$

3.2.3.4 *Performance of Cushion Devices*

(a) *Cushion forces*

After the cushion plunger has begun to penetrate into the head cover bore opening to the exhaust port, the working fluid in the cushion chamber is forced to flow into the pilot conduit and it passes through the cushion valve throttle, as shown in Fig. 3.39. Since the pilot conduit and the cushion valve throttle may be considered as choke throttles, the flow rate q passing through the cushion device, as well as the flow resistances R_1 and R_2 can be expressed as the following equations by using Eqs. (2.75) and (2.104)

$$q = \frac{p_1 - p_2}{R_1 + R_2} = Av, \tag{3.51}$$

$$R_1 = \frac{8\mu l_1}{\pi r^4}, \quad R_2 = \frac{12\mu l_2}{\pi d h^3}. \tag{3.52}$$

It is assumed that the leakage between the plunger and bore is negligible, and notations are used as follows:

μ: Viscosity, p_1: Pressure in cushion chamber, p_2: Pressure in head cover chamber, r: Radius of pilot conduit, l_1: Length of pilot conduit, h: Annular clearance in cushion valve throttle, d: Annular clearance diameter, l_2: Annular clearance length, R_1: Pilot conduit flow resistance, R_2: Flow resistance in annular clearance of valve throttle, A: Piston area, v: Piston speed.

Since the cushion force $f_{d1} = Ap_1 \approx A(p_1 - p_2)$ withstands the piston movement, the cushion force f_{d1} is defined by Eq. (3.53), recalling Eq. (3.51) and $q = Av$:

$$f_{d1} \approx A(p_1 - p_2) = A^2(R_1 + R_2)v = C_1v \left.\begin{array}{c} \\ \\ \end{array}\right\}.$$
$$C_1 = A^2(R_1 + R_2) \qquad\qquad (3.53)$$

Thus, the cushion force f_{d1} is in proportion to the piston speed v.

When the cushion valve throttle is regarded as an orifice and the flow resistance in the pilot conduit is negligibly small, the cushion force $f_{d2} = Ap_1$ withstanding the piston movement is given by Eq. (3.54), recalling Eq. (2.59) and $q = Av$

$$f_{d2} = Ap_1 \approx \frac{\rho A^3}{2C_d^2 A_0^2}v^2 = C_2v^2 \left.\begin{array}{c} \\ \\ \end{array}\right\},$$
$$C_2 = \frac{\rho A^3}{2C_d^2 A_0^2} \qquad\qquad (3.54)$$

where C_d: Flow coefficient of cushion orifice, A_0: Cushion orifice area. Equation (3.54) shows that the cushion force f_{d2} is proportional to the square of the piston speed.

(b) *Cushion performance on the basis of a choke throttle*

Let us consider the cushion performance under the condition when the cushion force $f_{d1} = C_1v$ withstands the inertia force of the equivalent mass m_e as shown in Eq. (3.50). The equation of motion of the piston in the cushion device is expressed as follows:

$$m_e\frac{dv}{dt} + C_1v = 0 \left.\begin{array}{c} \\ \\ \end{array}\right\},$$
$$v(0) = V_0 \qquad\qquad (3.55)$$

where the time t and the piston transfer distance x are defined respectively as $t = 0$ and $x = 0$ at just the time when the plunger has begun to penetrate into the head cover bore at the initial velocity $v(0) = V_0$. The solution of Eq. (3.55) is given as follows:

$$v = V_0 e^{-(c_1/m_e)t}.$$ (3.56)

The piston transfer distance x is derived recalling Eq. (3.56):

$$x = \int_0^t v\,dt = \frac{V_0 m_e}{C_1}[1 - e^{-(c_1/m_e)t}].$$ (3.57)

Recalling Eqs. (3.56) and (3.57), the piston velocity v is derived as follows:

$$v = V_0 - \frac{C_1}{m_e}x.$$ (3.58)

Substituting $v = 0$ into Eq. (3.58) gives the transfer distance x_m at the time when the piston stops

$$x_m = \frac{m_e V_0}{C_1}.$$ (3.59)

Substituting $v = V_0/10$ into Eq. (3.56), the time t_m at the velocity $v = V_0/10$ is obtained as follows:

$$t_m = 2.3\frac{m_e}{C_1} = 2.3\frac{x_m}{V_0}.$$ (3.60)

(c) *Cushion performance on the basis of an orifice throttle*

The equation of motion of the piston with the cushion device is expressed by Eq. (3.61) for the condition when the cushion force $f_{d2} = C_2 v^2$ withstands the inertia force of the equivalent mass m_e

$$\left.\begin{array}{c} m_e\dfrac{dv}{dt} + C_2 v^2 = 0 \\[2mm] v(0) = V_0 \end{array}\right\},$$ (3.61)

where $v(0) = V_0$ denotes the initial piston velocity at the time when the plunger has begun to penetrate into the cushion chamber. The cushion force $f_{d2} = C_2 v^2$ is arranged as $f_{d2} = C_2 v(dx/dt)$ by using the infinitesimal piston transfer distance dx during the infinitesimal time dt. Substituting

$C_2v^2 = C_2v(dx/dt)$ into Eq. (3.61) gives

$$m_e\frac{1}{v}dv = -C_2dx. \tag{3.62}$$

Integrating Eq. (3.62) in the area from the initial velocity $v = V_0$ at $x = 0$ to the velocity v at a transfer distance x, Eq. (3.62) is rearranged as follows:

$$m_e\int_{V_0}^{v}\frac{1}{v}dv = -C_2\int_{0}^{x}dx. \tag{3.63}$$

Therefore, the piston velocity v is obtained as follows:

$$v = V_0e^{-(c_2/m_e)x}. \tag{3.64}$$

Substituting $v = V_0/10$ into Eq. (3.64) yields the transfer distance x_m

$$x_m = 2.3\frac{m_e}{C_2}. \tag{3.65}$$

Substituting $v = dx/dt$ into Eq. (3.64) gives

$$e^{(C_2/m_e)x}dx = V_0dt. \tag{3.66}$$

By integrating Eq. (3.66)

$$\int_{0}^{x_m}e^{(c_2/m_e)x}dx = V_0\int_{0}^{t_m}dt. \tag{3.67}$$

Therefore, the time t_m at $x_m = 2.3(m_e/C_2)$ is obtained as follows:

$$t_m = 9\frac{m_e}{C_2V_0}. \tag{3.68}$$

3.3 Hydraulic Control Valves

Hydraulic control valves are classified as directional control valves, pressure control valves and flow control valves. The flow directions, flow rates and pressures in hydraulic systems are typically controlled by using either poppet valves or spool valves.

3.3.1 *Performance of Spool Valves and Poppet Valves*

The spool valve shown in Fig. 3.42(a) controls the flow by varying the area of the port that overlaps the sliding spool lands. A spool with multiple spool lands is able to control multiple port areas at the same time, and is suitable for continuous control of port areas. Its major drawbacks are susceptibility to fluid contaminants and the need for high precision processing.

As shown in Fig. 3.42(b), the poppet valve controls the flow by varying the port opening area, which consists of the clearance between the valve seat and the poppet cone. Poppet valves are suitable for use in on–off control valves because they provide tight sealing and good resistance to fluid contaminants, although they require a large force to operate them compared to spool valves. The flow rate Q through the valve port is denoted by applying Eq. (2.59)

$$Q = A_v u_1 = C_d A_v \sqrt{\frac{2\Delta p}{\rho}}, \qquad (3.69)$$

where u_1 is the mean velocity at the port, A_v is the opening port area, C_d is the flow coefficient of the port, ρ is the working fluid density, and Δp is the differential pressure between the upper and the lower flow at the port.

It is recommended that the power needed to operate the valve should be as small as possible compared to the controlled fluid power. In electro-hydraulic systems, servo valves actually control the fluid power by means of an electric power less than $1/10^5$ of the fluid power. However, the fluid flow forces produced by the controlled flow often act on the valve causing

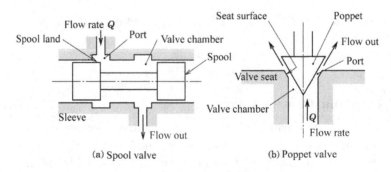

Fig. 3.42   Spool valve and poppet valve.

the valve performance to deteriorate. Let us begin with a discussion on the fluid flow forces acting on the valve in order to get a better insight into how the valve works.

3.3.1.1 *Axial Fluid Flow Forces and Moving Behavior of Spool*

(a) *Steady axial fluid flow force acting on spool*

Figure 3.43(a) illustrates the flow in a spool valve. While the working fluid is passing through the valve chamber, the change of fluid velocity gives rise to a pressure change in the spool chamber, which acts on the spool as a fluid flow force. Though it is difficult to determine the pressure distribution in the spool chamber, the axial fluid flow force $-F$ may be derived by applying the momentum theory outlined in Section 2.2.3. Since the spool chamber is regarded as a control volume, the axial force F_s acting on the spool may be derived from the reaction against fluid force $-F$ acting on the control surface. The axial spool force $F_s = F$ is defined recalling Eq. (2.54)

$$-F = \frac{d}{dt} \int_0^L \rho u_c A_c dL + \int_0^{A_v} \rho u^2 \cos(90° - \phi)\cos\phi dA_v, \tag{3.70}$$

where ϕ is the velocity angle of fluid flowing into the spool chamber, L is the distance between inlet and outlet port, u is the fluid velocity flowing into spool chamber, u_c is the mean axial fluid velocity in the spool chamber, A_v is the opening area of the spool control port and A_c is the cross-sectional area of the flow path in the spool chamber.

As long as the flow is steady and the pressure fluctuations are negligible, the first term on the right-hand side of Eq. (3.70) may be considered to be

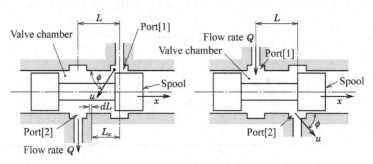

(a) Spool valve with inlet control port   (b) Spool valve with outlet control port

Fig. 3.43   Flow in the spool chamber.

zero. The second term in Eq. (3.70) expresses the axial momentum rate of the working fluid passing through the control volume, because the axial momentum rate of the fluid flowing out of the control volume may be taken to be zero. If the axial fluid velocity flowing into the spool chamber and the mass flow rate are denoted as $u_1 \cos\phi$ and ρQ using the mean fluid flow velocity u_1 at the port, then the axial momentum rate of the fluid flowing into spool chamber may be approximated to $\rho Q u_1 \cos\phi$ when the force acting in the right direction is assigned the plus sign. Therefore the steady axial force acting on the spool $F_s = F$ is approximated recalling $u_1 = \sqrt{2\Delta p/\rho}$ given in Eq. (2.58)

$$F_s = -\int_0^{A_v} \rho u^2 \cos(90° - \phi) \cos\phi \, dA_v \approx -\rho Q u_1 \cos\phi = -\rho Q \sqrt{\frac{2\Delta p}{\rho}} \cos\phi.$$

$$(3.71)$$

Equation (3.71) is rearranged recalling Eq. (3.69) as follows:

$$F_s = -2C_d A_v \Delta p \cos\phi. \qquad (3.72)$$

Figure 3.43(b) illustrates a spool valve that controls the flow at the outlet port [2] in the spool chamber. Applying the momentum theory to the flow in Fig. 3.43(b), we obtain the identical equation to Eq. (3.72) because the axial fluid velocity flowing out of the chamber has direction to the right whereas the mean axial fluid velocity flowing into the chamber is zero. Therefore, Eq. (3.72) is applicable to each of the cases shown in Figs. 3.43(a) and (b), and the axial steady forces acting on the spool are the stable restoring forces of resistance to the spool movement.

As shown in Fig. 3.44, the velocity angle ϕ of fluid flowing into or out of the spool chamber varies with the ratio of the port length, x, to the

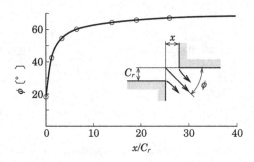

Fig. 3.44   Flow angle into spool chamber.

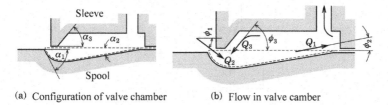

(a) Configuration of valve chamber   (b) Flow in valve camber

Fig. 3.45 Chamber configuration to reduce axial fluid force.

clearance between the spool and the spool bore, C_r. However, the angle ϕ tends to the value of 69° when the opening port length x is significantly larger than the small clearance C_r, which is commonly the case.

(b) *Countermeasures to reduce steady axial fluid force*

Equation (3.72) shows that the steady axial force increases as the opening port area and the differential pressure increase. The following three countermeasures (1), (2) and (3) can be taken to reduce the steady axial fluid force.

(1) *Modifying the configuration in the spool chamber*

Figure 3.45 illustrates a spool chamber configuration that reduces the axial steady fluid force. There are flow rates of Q_2 and Q_3 into the control volume enclosed with broken line in the spool chamber, and there is a flow rate Q_1 out of the control volume. Since the axial force acting on the spool is the reaction against the axial fluid flow force acting on the control surface, the steady axial force on the spool F_s is denoted recalling Eq. (3.71)

$$F_s = \rho Q_1 u_1 \cos \phi_1 - \rho Q_2 u_2 \cos \phi_2 + \rho Q_3 u_3 \cos \phi_3, \qquad (3.73)$$

where u_1, u_2 and u_3 are the mean fluid velocities corresponding to the flow rates Q_1, Q_2 and Q_3, respectively. By designing the chamber configuration angles α_1, α_2 and α_3 so as to reduce the force F_s in Eq. (3.73), the axial spool force may also be reduced.

(2) *Modifying the entrance configuration of the control port*

By modifying the entrance configuration of the control port as shown in Fig. 3.46, fluid velocities oppose each other at the spool port entrance, and the mean axial velocity flowing into the chamber is reduced. Consequently, the axial spool force is reduced.

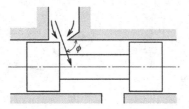

Fig. 3.46   Port configuration to reduce axial fluid force.

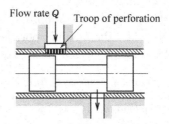

Fig. 3.47   Perforation troop port to reduce axial fluid force.

(3) *Using the port comprising several small holes*

Figure 3.47 illustrates the spool valve with the port composed of several small holes. When the port opens over a number of small holes, the fluid flows into the spool chamber in the direction of the holes, which is normal to the spool axis, so that the axial spool force is significantly reduced.

(c) *Dynamic characteristics of spool*

Let us now consider an unsteady fluid force $-F_t$ denoted by the first term in the right-hand side of the Eq. (3.70). This force is a consequence of unsteady flow in the spool chamber. The unsteady axial force on spool F_t is defined as follows:

$$F_t = -\frac{d}{dt}\int_0^L \rho u_c A_c dL = -\frac{d}{dt}\int_0^L \rho Q dL = -\rho L\frac{dQ}{dt}. \qquad (3.74)$$

The length of the opening port x corresponds to the spool transfer distance from the spool center position, and the area of the opening port is denoted as $A_v = wx$. Substituting the opening port area $A_v = wx$ and Eq. (3.69) into Eq. (3.74) gives

$$F_t = -LwC_d\sqrt{2\rho\Delta p}\frac{dx}{dt}. \qquad (3.75)$$

Equation (3.75) shows that the unsteady axial force F_t is a damping force proportional to the port opening velocity dx/dt whilst the spool is moving.

When the inlet control port in Fig. 3.43(a) is opened by a spool movement to the right ($dx/dt > 0$), the fluid in the spool chamber is accelerated in its flow to the left by an axial force whose magnitude depends on the increase of the flow rate differential dQ/dt. Since the unsteady axial force acting on the spool is the reaction force to the fluid flow in the spool chamber, the spool receives an unsteady axial force acting in the right direction. That is, the unsteady axial force prompts the spool movement to open the control port. In this case, it is called a minus damping force.

When the outlet port in Fig. 3.43(b) is opening due to a spool movement to the right, the spool chamber fluid is accelerated in its flow to the right by the unsteady axial fluid flow force related to the flow rate differential dQ/dt. That is, the unsteady axial force acting on the spool in Fig. 3.43(b) is a positive damping force withstanding the spool movement.

Figure 3.48 illustrates a four-way three-position flow control valve and the central spool land serves as the inlet control port of the two spool chambers in accordance with the spool movement direction. When the supply port opens as shown in Fig. 3.48, the working fluid flows into the left spool chamber opening to the actuator port A. Then, the working fluid in the right spool chamber is discharged to the reservoir.

The left and right spool chambers in Fig. 3.48 correspond to spool chambers shown in Figs. 3.43(a) and (b), respectively. Therefore, the axial spool force is derived as follows, applying the Eqs. (3.72) and (3.75) for the

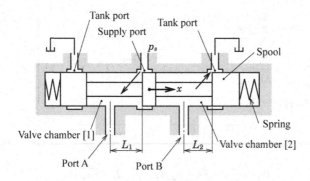

Fig. 3.48   Four-way three-position flow control valve.

left and the right spool chamber flows

$$F = -2C_d wx(\Delta p_1 + \Delta p_2)\cos\phi + C_d w(L_1\sqrt{2\rho\Delta p_1} - L_2\sqrt{2\rho\Delta p_2})\frac{dx}{dt},$$
(3.76)

where the suffix 1 and 2 correspond to the left and right spool chambers [1] and [2], respectively. When the spool is supported at the central position by the spring without any initial deflection, the equation of motion of the spool is given as follows:

$$F = m\frac{d^2x}{dt^2} + B_f\frac{dx}{dt} + kx,$$
(3.77)

where m is spool mass, k is spring constant and B_f is the fluid viscosity coefficient in the clearance between the spool and the bore.

Substituting Eq. (3.76) into Eq. (3.77) gives

$$\frac{d^2x}{dt^2} + \left[\frac{B_f + C_d w\sqrt{2\rho}(L_2\sqrt{\Delta p_2} - L_1\sqrt{\Delta p_1})}{m}\right]\frac{dx}{dt}$$
$$+ \left[\frac{2C_d w(\Delta p_1 + \Delta p_2)\cos\phi + k}{m}\right]x = 0$$
(3.78)

Applying Hurwitz stability criterion (see Section 4.2.5) to Eq. (3.78) gives the stability conditions for the spool behavior

$$B_f + C_d w\sqrt{2\rho}(L_2\sqrt{\Delta p_2} - L_1\sqrt{\Delta p_1}) > 0.$$
(3.79)

Unless the conditions implied in Eq. (3.79) are satisfied, minute perturbations will create large movements of the spool. The solution to Eq. (3.78) is obtained using the initial conditions i.e. $x(0) = 0$ and $dx/dt = V_0$ as follows:

$$x = \frac{V_0}{\omega_n\sqrt{\zeta^2-1}}e^{-\zeta\omega_n t}\sinh\omega_n\sqrt{\zeta^2-1}t \quad (\zeta > 1)$$

$$x = \frac{V_0}{\omega_n\sqrt{1-\zeta^2}}e^{-\zeta\omega_n t}\sin\omega_n\sqrt{1-\zeta^2}t \quad (\zeta < 1)$$

$$x = V_0 e^{-\omega_n t} \quad (\zeta = 1)$$

$$\omega_n = \sqrt{\frac{2C_d w(\Delta p_1 + \Delta p_2)\cos\phi + k}{m}}$$

$$\zeta = \frac{B_f + C_d w\sqrt{2\rho}(L_2\sqrt{\Delta p_2} - L_1\sqrt{\Delta p_1})}{2\sqrt{m}[2C_d w(\Delta p_1 + \Delta p_2)\cos\phi + k]^{1/2}}$$
(3.80)

where ω_n is undamped natural angular frequency and ζ is the damping coefficient.

3.3.1.2 *Axial Fluid Force and the Motion of the Poppet*

(a) *Steady axial force acting on the poppet*

Figure 3.49 illustrates the opening port area $A(x)$ of the poppet valve. The opening port area corresponding to the valve lift x is the annular surface area between the valve seat and the poppet cone. It is given by the relevant area shown in Fig. 3.49(b)

$$A(x) = \pi \left[\left(\frac{d_m}{2\cos\alpha} \right)^2 - \left(\frac{d_m}{2\cos\alpha} - x\sin\alpha \right)^2 \right] \cos\alpha$$

$$\approx \pi x \sin\alpha (d_m - x\sin\alpha\cos\alpha), \tag{3.81}$$

where 2α denotes the vertical angle of the poppet cone as shown in Fig. 3.49(a). The length $(d_m - x\sin\alpha\cos\alpha)$ may be approximated as d_m because the valve lift x is negligibly small compared with the diameter d_m. Using the approximation $d \approx (d_m - x\sin\alpha\cos\alpha)$, the opening port area $A(x)$ is expressed as follows:

$$A(x) \approx \pi x d \sin\alpha. \tag{3.82}$$

Figure 3.50(a) shows the flow direction in which the differential pressure Δp between the upper and the lower flow at the poppet acts to open the poppet valve, whereas the differential pressure Δp in Fig. 3.50(b) acts to close the poppet valve. The axial force associated with the differential

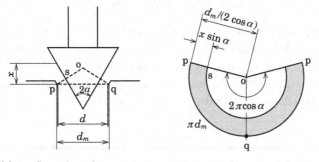

(a) Configuration of poppet port　　(b) Development of opening area

Fig. 3.49　Opening area of the poppet valve.

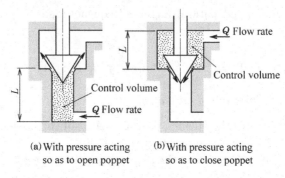

(a) With pressure acting   (b) With pressure acting
so as to open poppet   so as to close poppet

Fig. 3.50   Flow in poppet chamber.

pressure Δp is denoted as $S\Delta p$ recalling the effective area S of the poppet receiving the pressure.

Regarding the poppet chamber as a control volume, the axial steady fluid force in Figs. 3.50(a) and (b) is denoted as $\rho Q u_1 \cos \alpha$, which is the differential momentum rate between the flow out of and into the chambers. Therefore, the axial force F_s acting on the poppet is formulated recalling Eq. (3.71)

$$F_s = \pm[S\Delta p - \rho Q u_1 \cos \alpha] = \pm[S\Delta p - Q\sqrt{2p\Delta p}\cos \alpha]$$
$$= \pm[S\Delta p - \rho C_d A(x)u_1^2 \cos \alpha] = \pm[S\Delta p - \pi C_d x d\Delta p \sin 2\alpha], \quad (3.83)$$

where the force acting in the direction to open the poppet valve is defined as plus, and the sign symbol $\pm$ corresponds to the flow in Figs. 3.50(a) and (b), respectively.

(b) *Dynamic characteristics of the poppet valve*

The unsteady axial fluid force in both Figs. 3.50(a) and (b) is denoted as $-\rho L dQ/dt$, with reference to Eq. (3.74). However, in Fig. 3.50(a) it acts to close the poppet whereas in Fig. 3.50(b) it acts to open the poppet. Therefore, the unsteady axial fluid force in Figs. 3.50(a) and (b) is obtained as follows, recalling Eqs. (3.69), (3.74) and (3.82)

$$F_t = -\rho L \frac{dQ}{dt} = \pm\left(-\pi L C_d d \sin \alpha \sqrt{2\rho\Delta p}\frac{dx}{dt}\right). \quad (3.84)$$

The equation of motion of the poppet is given as follows for the case when the poppet valve is supported by a spring with a constant k and a damper with the viscosity coefficient B_f

$$m\frac{d^2x}{dt^2} + B_f\frac{dx}{dt} + k(x + x_0) = F_s + F_t, \qquad (3.85)$$

where x_0 is the initial deflection of the spring, and m is the mass of the poppet valve. Substituting Eqs. (3.83) and (3.84) into Eq. (3.85) gives the equations of motion for the poppet. For the flow shown in Fig. 3.50(a):

$$\frac{d^2x}{dt^2} + \left(\frac{B_f + \pi\sqrt{2\rho\Delta p}C_d Ld\sin\alpha}{m}\right)\frac{dx}{dt}$$

$$+ \left(\frac{k + \pi C_d d\Delta p\sin 2\alpha}{m}\right)x + \frac{kx_0 - S\Delta p}{m} = 0. \qquad (3.86)$$

For the flow direction shown in Fig. 3.50(b):

$$\frac{d^2x}{dt^2} + \left(\frac{B_f - \pi\sqrt{2\rho\Delta p}C_d Ld\sin\alpha}{m}\right)\frac{dx}{dt}$$

$$+ \left(\frac{k - \pi C_d d\Delta p\sin 2\alpha}{m}\right)x + \frac{kx_0 + S\Delta p}{m} = 0. \qquad (3.87)$$

Applying the Hurwitz stability criterion (see Section 4.2.5) to Eq. (3.87) yields a conclusion that the dynamic poppet behaviors denoted by Fig. 3.50 (b) are unstable unless the following Eq. (3.88) can be satisfied

$$B_f > \pi C_d Ld\sin\alpha\sqrt{2\rho\Delta p}. \qquad (3.88)$$

3.3.2 *Pressure Control Valves*

Pressure control valve is the term used for valves with the function of pressure control, such as pressure relief, pressure reducing, sequential operation, counterbalance to maintain back-pressure, pressure switch, etc.

3.3.2.1 *Relief Valve*

A relief valve may be set up near a pump outlet to limit the system pressure within a specified range. Figure 3.51 illustrates the direct relief valve, in which the spring force acts to press the poppet against the valve seat. When the system pressure approaches the pressure limit, the poppet port begins to open, because the axial force caused by the system pressure overcomes the spring force, and then a part of the pump displacement returns to the reservoir through the poppet valve. As the line pressure rises, the flow rate through the relief valve increases, and eventually all of the pump

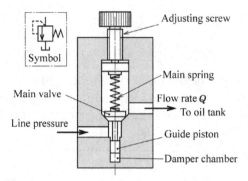

Fig. 3.51   Direct relief valve.

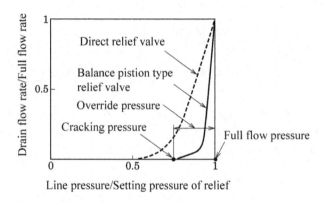

Fig. 3.52   Characteristics of relief valve.

displacement returns to the reservoir through the poppet valve at the setting pressure. Then the actuators in the system maintain the pressure at
the maximum limit. Figure 3.52 shows the flow characteristics in the relief
valve. The line pressure that starts to open the relief valve is called the
cracking pressure, while the line pressure at which all of the pump displacement flows through the relief valve is defined as the full flow pressure,
and the pressure difference between the full flow pressure and the cracking
pressure is called the override pressure. It is recommended that the over-
ride should be as small as possible, otherwise energy losses in the system
increase significantly.

The static force balance at the poppet is formulated recalling Eq. (3.83)

$$Sp_s - Q\sqrt{2\rho p_s}\cos\alpha = k(x + x_0), \tag{3.89}$$

where p_s is the line pressure and x_0 is the initial spring deflection. Substituting $Q = 0$ and $x = 0$ into Eq. (3.89) gives the cracking pressure p_c as follows:

$$p_c = \frac{kx_0}{S}. \tag{3.90}$$

Substituting Eq. (3.82) into Eq. (3.69) gives the flow rate Q through the poppet port

$$Q = \pi C_d x d \sin \alpha \sqrt{\frac{2p_s}{\rho}}. \tag{3.91}$$

Substituting Eq. (3.91) into Eq. (3.89) gives

$$p_s = \frac{k(x + x_0)}{S - \pi C_d x d \sin 2\alpha}. \tag{3.92}$$

Equation (3.92) shows that the over-ride $p_s - p_c$ becomes nearly zero when the design requirements implied in Eq. (3.93) are satisfied

$$\left. \begin{array}{c} x \ll x_o \\ \pi C_d x d \sin 2\alpha \ll S \end{array} \right\}. \tag{3.93}$$

However, it is difficult to design the direct relief valve that satisfies the design conditions because they demand excessively large spring dimensions, due to the large spring load Sp_s. Therefore this type of relief valve has to suffer a large over-ride performance. Moreover, the large spring load gives rise to valve chattering in the neighborhood of the cracking pressure unless the damper device is provided with a poppet. Chattering occurs when the force lifting the poppet off is in excess of the theoretical cracking spring force due to the friction effects of the poppet valve, and working fluid creeps through the small clearance in the valve seat generating a force to open the poppet valve. The poppet valve then tends to jump open, creating a subsequent fall in pressure such that the poppet rapidly resets itself. Repetition of this series of events causes valve chatter.

Figure 3.53 illustrates the improved direct relief valve in which the over-ride performance is vastly improved by using a differential piston type poppet to make the spring load small.

Figure 3.54 illustrates a balance piston type relief valve that has excellent over ride and precise pressure regulation performance. The line flow is piloted through a throttle to the pilot poppet chamber opening to the

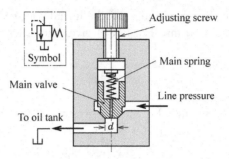

Fig. 3.53   Improved direct relief valve.

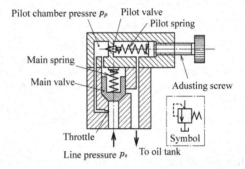

Fig. 3.54   Balance piston type relief valve.

upper chamber of the main poppet. When the line pressure p_s is smaller than the limiting pressure set by the initial deflection of the pilot spring, the main poppet closes due to the action of the main spring force because the pilot poppet chamber pressure p_p is equal to the line pressure p_s.

When the line pressure p_s rises up to the cracking pressure in the poppet chamber, the pilot poppet port opens slightly and the working fluid in the poppet chamber flows into the drain conduit. Thus the pressure in the pilot poppet chamber falls, and consequently, the main poppet opens, thereby regulating the line pressure so that the pressure limit is not exceeded. The balance piston type relief valve has a superior override performance because it can be designed so as to give a small spring load force by making a pilot poppet chamber. The flow rate Q_m passing through the main poppet port is obtained recalling Eq. (3.69) and Eq. (3.82)

$$Q_m = \pi C_{dm} d_m x_m \sin \alpha_m \sqrt{\frac{2p_s}{\rho}}, \qquad (3.94)$$

The static force balance in the main poppet is formulated as follows:

$$S_m(p_s - p_p) - \rho Q_m \sqrt{\frac{2p_s}{\rho}} \cos \alpha_m = k_m(x_m + x_{0m}), \qquad (3.95)$$

where the suffixes m and p correspond to the main poppet and pilot poppet, respectively, and the second term in the left-hand side of Eq. (3.95) stands for the reaction of the steady fluid flow force in the poppet chamber.

The flow rates Q_{p1} flowing into the pilot chamber through the throttle and the flow rate Q_{p2} flowing through pilot poppet are expressed as follows:

$$Q_{p1} = C_{d1} a \sqrt{\frac{2(p_s - p_p)}{\rho}}, \qquad (3.96)$$

$$Q_{p2} = \pi C_{d2} x_p d_p \sqrt{\frac{2p_p}{\rho}} \sin \alpha_p, \qquad (3.97)$$

where a and x_p denote the throttle area and the lift of the pilot poppet, respectively, and C_{d1} and C_{d2} are the flow coefficients of the throttle and the pilot poppet port, respectively. Since the flow rate Q_{p1} in Eq. (3.96) equals the flow rate Q_{p2} in Eq. (3.97) during steady flow, the relationship between the line pressure p_s and the pilot chamber pressure p_p is derived recalling that $Q = Q_{p1} = Q_{p2}$:

$$p_s = \left[1 + \left(\frac{\pi C_{d2} x_p d_p \sin \alpha_p}{C_{d1} a} \right)^2 \right] p_p. \qquad (3.98)$$

While the static force balance in the pilot poppet is formulated as follows:

$$S_p p_p - \rho Q_{p2} \sqrt{\frac{2p_p}{\rho}} \cos \alpha_p = k_p(x_p + x_{0p}). \qquad (3.99)$$

Substituting Eq. (3.97) into Eq. (3.99) gives

$$p_p = \frac{k_p(x_p + x_{0p})}{S_p - \pi C_{d2} x_p d_p \sin 2\alpha_p}. \qquad (3.100)$$

Substituting Eq. (3.100) into Eq. (3.98) gives

$$p_s = \frac{k_p(x_p + x_{0p})}{S_p - \pi C_{d2} x_p d_p \sin 2\alpha_p} \left[1 + \left(\frac{\pi C_{d2} x_p d_p \sin \alpha_p}{C_{d1} a} \right)^2 \right]. \qquad (3.101)$$

Equations (3.100) and (3.101) show that both the line pressure p_s and the pilot chamber pressure p_p approach the cracking pressure $p_c = k_p x_0 / S_p$

when the design conditions implied by Eq. (3.102) are satisfied:

$$S_p \gg \pi C_d x_p d_p \sin 2\alpha_p, \quad x_p \ll x_{0p}, \quad \left(\frac{\pi C_{d2} d_p x_p \sin \alpha_p}{C_{d1} a}\right)^2 \ll 1. \quad (3.102)$$

Then, the line pressures p_s are regulated by the deflection of the pilot spring, and the override becomes nearly zero. It is easy to satisfy the design conditions for the balance piston type relief valve because the spring load force $S_p p_p$ is small compared to the spring load force $S p_s$ in direct spring valve types.

3.3.2.2 *Pressure reducing Valves*

Pressure reducing valves are positioned in the line between the main circuit and branch circuits in order to restrict the pressures in the branch circuits to a specified reduced level. Figure 3.55 illustrates a balance piston type pressure reducing valve. The main circuit flow is bypassed to the branch circuit through the spool port, and the flow in the branch circuit is piloted to the pilot chamber through the throttle A. When the pilot poppet port closes under the action of the pilot spring force, the spool control port opens due to the main spring force, so that the branch circuit pressure almost equals the main circuit pressure. When the pilot spring force is reduced by regulating the manual screw, the pilot poppet opens and the pressure p_p in the pilot chamber is reduced. Thereby, the spool moves up so as to decrease the spool port area, with the result that the pressure p_2 in the branch circuit is reduced. Since the reduced branch circuit pressure p_2 acts on the lower surfaces of the spool, the spool stops at the position where

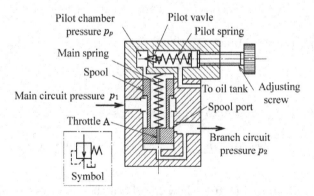

Fig. 3.55　Pressure reducing valve.

the forces acting on the upper and lower surface of the spool are balanced. Assuming that the main spring force is negligibly small compared with the spool force dependent on the pressure p_2, the branch pressure p_2 is regulated to reduce pressure p_p set by the manual regulating screw.

The flow rate Q_{p1} through the throttle A and the flow rate Q_{p2} through the pilot poppet are expressed as follows:

$$Q_{p1} = C_{d1} a \sqrt{\frac{2(p_2 - p_p)}{\rho}}, \qquad (3.103)$$

$$Q_{p2} = \pi C_{d2} d_p x_p \sin \alpha_p \sqrt{\frac{2p_p}{\rho}}, \qquad (3.104)$$

where a denotes the area of the throttle A shown in Fig. 3.55. The static force balance in the pilot poppet is formulated as follows:

$$S_p p_p - \rho Q_{p2} \sqrt{\frac{2p_p}{\rho}} \cos \alpha_p = k_p (x_p + x_{0p}). \qquad (3.105)$$

Equations (3.103) and (3.104) correspond to Eqs. (3.96) and (3.97), respectively and Eq. (3.105) correspond to Eq. (3.99). Therefore, Eqs. (3.100) and (3.101) applicable to the relief valve will also apply to reducing valves shown in Fig. 3.55. Accordingly, Eq. (3.106) can be derived as long as the design condition implied by Eq. (3.102) is satisfied

$$p_p \approx p_2 \approx \frac{k_p x_{0p}}{S_p}. \qquad (3.106)$$

Equation (3.106) shows that the branch pressure p_2 can be regulated by the deflection of the pilot spring.

3.3.2.3 *Unloading Valves*

An unloading valve is a pressure control valve used to unload the pressure line by diverting the flow in the pressure line to a reservoir, dependent on the specified pilot pressure. Figure 3.56 illustrates an unloading valve in which the spool moves up when the pilot pressure exceeds the pressure corresponding to the spring force preset by a manual regulating screw. The spool movement in the upward direction makes the spool port open, so that the line pressure is unloaded. Figure 3.57 illustrates a typical circuit incorporating an unloading valve. When the pilot pressure associated with the actuator load is less than the setting pressure, the actuator in the system moves at high speed because the two pumps supply the fluid into

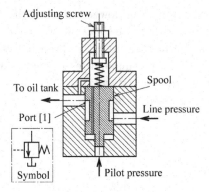

Fig. 3.56   Unloading valve.

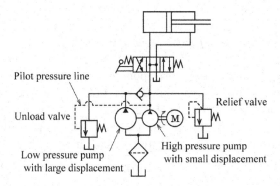

Fig. 3.57   Typical unloading circuit.

the actuator. However, the hydraulic circuit achieves some power saving by unloading the low pressure pump as long as the pilot pressure exceeds the setting pilot pressure.

3.3.2.4 *Sequence Valves*

A sequence valve is a device that manages the working procedure of actuators in a hydraulic circuit. Figure 3.58 illustrates a sequence valve whose spool port [1] closes during the first stage when the actuator is working within the pressure preset by a manual adjusting screw. When the first stage actuator stops, the first stage line pressure exceeds the preset pressure. Then the spool port [1] of the sequence valve opens, and consequently the actuator in the second stage line begins to run by diverting the first stage flow to the second stage line.

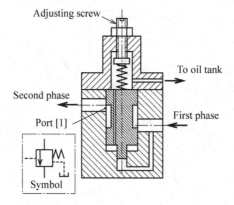

Fig. 3.58 Sequence valve.

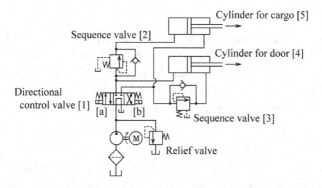

Fig. 3.59 Application circuit for sequence valve.

Figure 3.59 shows an example of a sequence circuit using sequence valves [2] and [3] with a check valve. If the four-way three-position directional flow control valve takes the [a] side block flow paths in the valve symbol, then the working fluid flows only into the left side chamber of the door cylinder [4] through the directional control valve [1] because the circuit pressure is within the preset pressure of the sequence valve [2]. When the cylinder [4] stops because of the fully opened door, the pressure in the circuit increases until it exceeds the preset pressure of the sequence valves [2] so that the working fluid flows into the left side chamber of the cylinder [5] to transfer a cargo.

If the four-way three-position directional flow control valve acts to take the [b] side block flow paths, then the working fluid flows only into the right

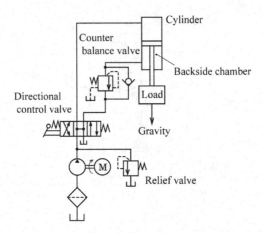

Fig. 3.60   Application circuit for counter balance valve.

side chamber of the cylinder for cargo [5] to return the piston to the original position, because the line pressure is below the preset pressure of sequence valve [3]. After the cylinder [5] has returned to its original position, the cylinder for door [4] acts to close the door because the circuit pressure rises and exceeds the preset pressure of sequence valve [3].

3.3.2.5 *Counter Balance Valves*

A counter balance valve has a similar configuration as a sequence valve as shown in Fig. 3.58, and it is used to control the speed of a vertical cylinder so that it should not fall freely due to gravity. As shown in Fig. 3.60, the working fluid in the backside chamber of the cylinder returns to the reservoir through the counter balance valve whose throttle area varies in accordance with the pressure in the back side line. Consequently, the counter balance valve controls the flow rate in the tank line in order to transfer the load at an adequate fall speed by the valve preset pressure.

3.3.3 *Flow Control Valves*

A flow control valve controls the flow rate in a hydraulic circuit, and is used as a speed control device of an actuator. Although the control of flow rate achieved by variable displacement pumps is superior in terms of energy efficiency, it is difficult to achieve a flow control with good response in the branch circuit. Flow control valves are suitable for controlling the

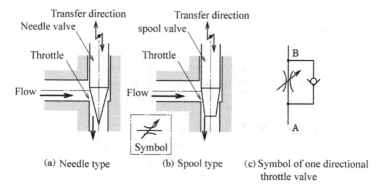

Fig. 3.61   Throttle valve.

actuator speed as part of the working cycle in hydraulic systems, and may be classified as throttle valves and pressure compensated flow control valves.

3.3.3.1 *Throttle Valves*

The flow rate passing through the throttle depends on the throttle area and the differential pressure between the upper and the lower flow of the throttle as shown in Eq. (2.59) and Fig. 2.14. Since a throttle valve regulates the flow rate by varying the throttle opening area, it should be used as a rough speed control when the running condition is such that the system pressure is almost constant. Figures 3.61(a) and (b) show the configurations of a needle type and a spool type throttles, respectively. In addition, Fig. 3.61(c) shows a symbol of a one directional throttle device comprising of a throttle and a check valve set in parallel with the throttle. This configuration is frequently used in hydraulic systems. It controls the flow rate in one direction of flow by regulating the throttle area, but it gives free flow in the opposite direction.

3.3.3.2 *Pressure Compensated Flow Control Valve*

The flow rate passing through a pressure compensated flow control valve is kept nearly constant at a pre-set level even if the line pressures change significantly. Figure 3.62 illustrates a pressure compensated flow control valve in which a spring is set between spool [1] and spool [2], and the supplied flow is piloted to the right end of spool [2] to control the area of the spool port B. The flow rate passing through the valve is regulated by setting the triangular throttle area of port A. The working fluid flows into the

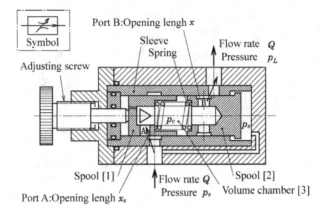

Fig. 3.62   Pressure compensated flow control valve.

valve chamber [3] between both spools through port A, and then flows out of port B. If the load pressure or the supply pressure suddenly changes, then the spool [2] moves left or right depending on the differential pressure. Thereby, the throttle area of port B varies to maintain the constant differential pressure at port A. Consequently the flow rate passing through port A can be kept constant regardless of the line pressure changes.

Regarding the spool positions when ports A and B are just beginning to open as the reference positions, the static force balance equation on the spool [2] is formulated as follows:

$$A_p(p_s - p_c) + \rho Q_A u_A \cos \phi_A + \rho Q_B u_B \cos \phi_B = k(x_0 + x_s - x), \quad (3.107)$$

where x_s is the displacement of spool [1], x is the displacement of the spool [2], $a(x_s)$ is the opening area of the port A, $b(x)$ is the opening area of control port B, u_A is the mean fluid velocity at port A, u_B is the mean fluid velocity at port B, $u_A \cos \phi_A$ is the axial fluid velocity element at port A, $u_B \cos \phi_B$ is the axial fluid velocity element at the port B, A_p is the spool area receiving axial pressure, x_0 is the spring deflection at a reference location, p_s is the supply pressure, p_c is the spool chamber pressure, p_L is the load line pressure, and ρ is the working fluid density.

The second and third terms in the left side of Eq. (3.107) denote the axial force generated by the steady flow passing through the valve, recalling Eq. (3.72). The flow rate Q_A passing through port A is expressed recalling

Eq. (3.69)

$$Q_A = C_{dA}a(x_s)\sqrt{\frac{2(p_s - p_c)}{\rho}}. \tag{3.108}$$

Recalling the continuity Eq. (2.42) in steady flow, the flow rate Q_B passing through port B equals the flow rate Q_A

$$Q = Q_A = Q_B = C_{dB}b(x)\sqrt{\frac{2(p_c - p_L)}{\rho}}. \tag{3.109}$$

The mean velocities u_A and u_B are described as follows:

$$u_A = \frac{Q_A}{a(x_s)}, \quad u_B = \frac{Q_B}{b(x)}. \tag{3.110}$$

Substituting Eqs. (3.108)–(3.110) into Eq. (3.107) gives

$$Q = C_{dA}a(x_s)\sqrt{\frac{2kx_0}{\rho A_p}\left(1 + \frac{x_s - x}{x_0}\right)}$$
$$\times \left\{1 + \frac{2C_{dA}^2 a(x_s)\cos\phi_A}{A_p} + \frac{2\left[C_{dA}a(x_s)\right]^2 \cos\phi_B}{b(x)A_p}\right\}^{-1/2}. \tag{3.111}$$

Therefore, the flow rate Q passing through the flow control valve depends on the opening area $a(x_s)$ regardless of the pressures p_s and p_L when the designing conditions implied in Eq. (3.112) are satisfied

$$\frac{x_s - x}{x_0} \ll 1, \quad \frac{2C_{dA}^2 a(x_s)\cos\phi_A}{A_p} + \frac{2\left[C_{dA}a(x_s)\right]^2 \cos\phi_B}{b(x)A_p} \ll 1. \tag{3.112}$$

3.3.3.3 *Flow Dividing Valve*

A flow dividing valve distributes the flow into two or more circuits on the basis of a specified flow rate ratio. Figure 3.63 illustrates a flow dividing valve to provide a constant ratio Q_1/Q_2 of the flow rate Q_1 passing through the orifice [1] to the flow rate Q_2 passing through the orifice [2]. The fluids passing through orifices [1] and [2] flow into the A-branch circuit and B-branch circuit, respectively, and the ratio of flow rates is denoted as

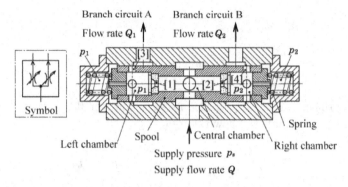

Fig. 3.63   Flow dividing valve.

follows:

$$\frac{Q_1}{Q_2} = \frac{C_{d1}a_1\sqrt{p_s - p_1}}{C_{d2}a_2\sqrt{p_s - p_2}},\qquad\qquad(3.113)$$

where a_1 and a_2 are the areas of fixed orifice [1] and [2], respectively, p_1 and p_2 are the pressures in the left and right spool chamber, respectively, as well C_{d1} and C_{d2} are flow coefficients in the orifice [1] and [2], respectively.

Let us consider the case when the working fluid flows into the flow dividing valve at a specified supply pressure p_s. By adjusting the spool position, the opening area of each variable orifices [3] and [4] in the left and right chamber may be regulated, thus making pressures p_1 and p_2 equal. Consequently, the flow rate ratio Q_1/Q_2 may be defined as $C_{d1}a_1/(C_{d2}a_2)$.

For instance, if the pressure p_1 increases due to a change in the A-branch circuit pressure p_{LA}, the differential pressure $p_1 - p_2$ causes the spool to move to the right to increase the opening area of the variable orifice [3] and coincidentally to decrease the opening area of the variable orifice [4]. Thereby, the increase of the pressure p_1 is suppressed and the pressure p_2 tends to increase, and consequently the spool stops at the position where differential pressure $p_1 - p_2$ becomes zero. Therefore the flow rate ratio Q_1/Q_2 is controlled to keep it constant regardless of the change of load pressure.

3.3.3.4 *Priority Control Valve*

A priority control valve first supplies working fluid to the priority circuit at the required flow rate, and then diverts any surplus fluid to the secondary circuit. Figure 3.64 shows a priority control valve. The fluid flows at first

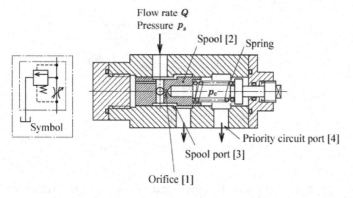

Fig. 3.64 Priority control valve.

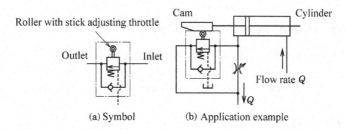

(a) Symbol   (b) Application example

Fig. 3.65 Deceleration valve.

into the priority circuit through orifice [1] at a flow rate determined by the difference in pressure $p_s - p_c$ between the upper and the lower flow of the orifice [1]. If the flow rate flowing into the priority circuit satisfies the demand then spool [2] moves to the right, because the spool force, which depends on the differential pressure $p_s - p_c$, overcomes the spring force. Consequently, surplus working fluid flows into the secondary circuit through the spool port [3].

3.3.3.5 *Deceleration Valve*

A deceleration valve regulates the flow rate by varying the throttle area in accordance with the movement of a cam. Figure 3.65(a) shows the graphic symbol of a deceleration device with a check valve in the bypath line in order to give the free flow path in reverse flow. Figure 3.65(b) illustrates a typical example of the control of cylinder speed using a deceleration valve. When the cylinder piston moves left, the cam pushes the roller down and the

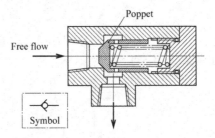

Fig. 3.66   Angle type check valve.

cylinder speed is reduced by the deceleration valve. In the right movement of the piston, it transfers at high speed because the fluid flow takes a free flow path through the check valve set in parallel to the deceleration valve.

3.3.4 *Directional Flow Control Valves*

Directional flow control valve is a general term for hydraulic valves that control the flow direction in hydraulic systems, such as check valves, shuttle valves and directional control valves managing the flow paths.

3.3.4.1 *Check Valve*

Figure 3.66 illustrates an angle type check valve in which the poppet is pressed against the valve seat by a small spring force. The working fluid flows freely through the check valve in the free flow direction because the fluid force overcomes the spring force. However, the reverse flow is blocked by the poppet due to the reverse flow pressure. The seal of the check valve in the reverse flow direction is so tight that the inner leakage is nearly zero even in large check valves, and the cracking pressure in free flow is so small that the pressure loss is negligible. There are also pilot type check valves in which the port is closed by a piloted pressure.

3.3.4.2 *Shuttle Valve*

A shuttle valve selects either the low or the high pressure circuit depending on the difference in pressure. Figure 3.67 illustrates a shuttle valve selecting the high pressure circuit. In this example, the high pressure circuit opens to the outlet port [3] and the lower pressure port closes, because the ball in the shuttle valve is pressed against the seat of the lower circuit port due to the differential pressure between the low and the high pressure circuit. In other

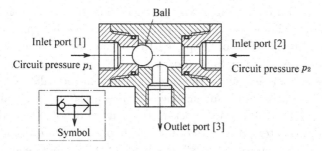

Fig. 3.67   Shuttle valve.

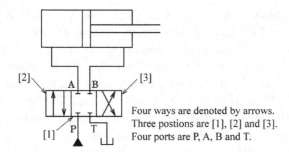

Four ways are denoted by arrows.
Three postions are [1], [2] and [3].
Four ports are P, A, B and T.

Fig. 3.68   Symbol of four-way three-position directional control valve.

words, if the pressure p_2 is higher than the pressure p_1, then the working fluid in the circuit [2] flows into the circuit [3]. While, if the pressure p_1 is higher than the pressure p_2, then the working fluid in the circuit [1] flows into the circuit [3]. Therefore, a shuttle valve can be used to select either the normal or the emergency circuit in a redundant hydraulic system, to protect against a pump failure.

3.3.4.3 *Directional Control Valve*

A directional control valve manages the flow directions by alternating spool positions as already described in Section 1.1. The terms 'way' and 'position' in a directional control valve are used to denote the flow path in a valve and the spool position in the valve, respectively. There are various directional control valves with a different number of ways and positions.

(a) *Four-way three-position directional control valve*

A four-way three-position directional control valve is a typical directional control valve to run a double acting actuator. Figure 3.68 illustrates the graphic symbol used for a four-way three-position directional control valve piping to the cylinder ports and the hydraulic source system. It has four ports consisting of: the port P piping to the fluid power source, the port T piping to the reservoir tank and the ports A and B piping to cylinder ports. The spool takes one of the three positions in the valve by sliding in spool bore, and the traffic arrangements between the ports are altered by the spool position. A central block in the symbol denotes that the spool locates at the center position, so that the spool lands close to all ports P, A, B and T. The arrows in the left side block of the symbol denote the flow directions when the spool shifts to the extreme right position as shown in Fig. 1.7(c).

The two parallel arrows denote that port P opens to port A and port T opens to port B. The right side block in the symbol with two crossed arrows denotes that port P and port T open to port B and port A, respectively when the spool shifts to the extreme left position. Table 3.8 shows four-way three-position directional control valves operating in a variety of modes that are possible with the proper design of the spool land. The center spool mode closing all of the ports is called "all ports close", while the center spool mode opening all of the port is called an "all ports open".

The functions of various directional control valves with different ways and positions can be denoted by the graphic symbols similar to the four-way three-position directional control valve symbols. Figure 3.69 shows the symbol of a one-way two-position valve called a shut off valve, which has the function of closing or opening a flow path port in the valve by a change in the spool position.

Four-way three-position directional control valves are divided into the sliding spool type, poppet type and rotary spool type, depending on their construction. The sliding spool type is the most popular because its operation requires little power and it has a compact structure.

As shown in Fig. 3.70, a poppet type four-way directional control valve consists of four poppet valves managing the flow path in the valve. The ports A and B are respectively piped to each port of a double acting actuator, and ports P and T are piped to the pressure line and the tank line, respectively. When poppet ports V_A, V_B, V_{TA} and V_{TB}, are closed, the flow passing through ports A, B, P and T is blocked. When the poppet valves V_A and V_{TB} coincidently open due to a pilot pressure controlled by a solenoid valve, port P opens to port A and port T opens to port B. When the poppet valves V_B and V_{TA} open concurrently, port P opens to port B and port T opens to port A. Since a poppet type control valve

Table 3.8   Port connections in central spool position.

Spool mode	Symbol	Explanation of function
All ports close	A B P T	All ports close when the spool is centred. When the spool position is altered, fluidic shock tends to occur because fluid flow is blocked.
All ports open	A B P T	Pump output is unloaded and the actuator is a floating condition at the center position of the spool. Fluidic shock is small when the spool position alters.
All ports open through throttle	A B P T	Fluidic shock is small when the spool position alters. This type is used as two position direction control valve.
ABT connections	A B P T	Pump output keeps a pressure and the actuator is in a floating condition at the center position of the spool.
PAT connections	A B P T	At the center position of the spool, pump output is unloaded and the actuator stops by means of blocking port B and directing supply fluid to port A.
PAB connections	A B P T	Port P is connected to port A and port B coincidentally at the center position of the spool. The PAB connections can be used for a construction of a differential pressure driving circuit.
PT connections	A B P T	Pump output is unloaded and the actuator keeps the position. The PT connections can be used for a construction of a tandem circuit.

(a) Normal open type   (b) Normal closed type

Fig. 3.69   Symbols of shut off valve.

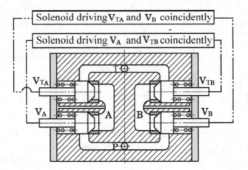

Fig. 3.70   Poppet type four-way directional control valve.

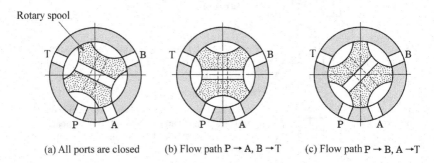

(a) All ports are closed   (b) Flow path P → A, B → T   (c) Flow path P → B, A → T

Fig. 3.71   Rotary spool type four-way directional control valve.

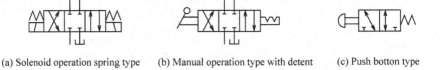

(a) Solenoid operation spring type   (b) Manual operation type with detent   (c) Push botton type

Fig. 3.72   Examples of valve symbols denoting the operating principles.

provides a tight seal when closed, it is suitable for use in on–off control valves, although it requires some mechanism for providing the necessary static pressure balance.

   Figure 3.71 illustrates the rotary spool type, four-way three-position directional control valve. The rotary spool closes all ports P, A, B and T at the central position as shown in Fig. 3.72(a). Rotating the spool counter-clockwise, ports P and T open to port A and B, respectively as shown in Fig. 3.71(b). Rotating the spool clockwise, port P and T open to

Table 3.9   Valve operation symbols.

Operation	Symbol				
Manual	Lever	Pedal	Twin pedal	Button	Push button
Mechanical	Spring	Roller	Plunger	Detent	
Electric	Single acting solenoid		Double acting solenoid		Variable magnetic actator
Pilot	Hydraulic pilot (Pressurized operation)		Hydraulic pilot (Pressure reducing operation)		Electric-hydraulic pilot

port B and A, respectively as shown in Fig. 3.71(c). The rotary spool type directional control valve is used only in low pressure hydraulic systems because the inner leakage performance is inferior to other types.

(b) *Valve operation*

The operating principles of valves are categorized as manual, mechanical, electric, oil or air piloted operation and some combinations of these. The graphic symbols for the valve operation types are shown in Table 3.9, and they are joined to the extreme block ends in the valve symbols as shown in Fig. 3.72. Figure 3.73 shows the typical frame of a four-way three-position directional control valve with a manual lever. The manual lever type directional control valve is frequently used in hydraulic control systems in construction machines because it is suitable for operation when the load movement has to be monitored.

Mechanical components such as rollers and plungers operate the valve by using a cam set up to the actuator. Electromagnetic operation operates the valve using a plunger type of solenoid as shown in Fig. 3.74. The magnetic force generated by the solenoid is sufficient to operate the valve in a short stroke because it is in inverse proportion to the square of the clearance between the magnet and the plunger, as expressed by Eq. (3.114)

$$F = \frac{1}{2}\varepsilon S \frac{(Ni)^2}{x^2},$$

(3.114)

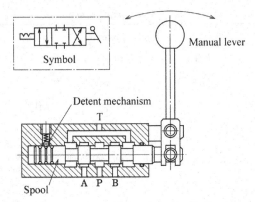

Fig. 3.73   Manual operated type four-way three-position directional control valve.

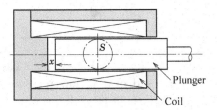

Fig. 3.74   Plunger type solenoid.

where S is the plunger cross-section area, x is the clearance between magnet and plunger, N is the coiling number, i is the electric current and ε is the magnetic permeability.

It is generally applied in sliding spool valves with the capacity less than 80 L/min because the force needed to operate a large spool gives rise to some problems. These include making the solenoid very large, but also generating a large shock force at the time when the spool transference has been complete. Figure 3.75 shows a solenoid operating a directional control valve that realizes the operation of the valve with a capacity of 80 L/min by using an electric circuit to moderate the shock involved. In directional flow control valves with the capacity exceeding 80 L/min a two stage solenoid-pilot operated device may be used. The main spool is operated by means of the piloted hydraulic pressure controlled by the small solenoid operation valve in the first stage. The solenoid-pilot operated valves are standardized for directional control valves designed to handle flow rates not exceeding 12000 L/min.

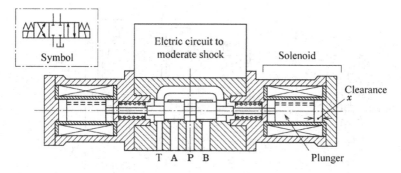

Fig. 3.75 Solenoid operation type four-way three-position control valve.

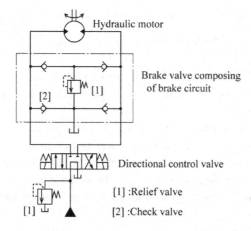

Fig. 3.76 Mixed valve constituting brake circuit.

Although the operation time of low or middle speed solenoid operated valves is in general about 0.1 s to 1.0 s, small-sized high speed solenoid valves can have operation times ranging from 0.5 ms to 10 ms, and are used in hydraulic systems in the construction machinery.

3.3.5 *Mixed Valves*

Control valves with a circuit function integrating the various valves are called mixed valves. They are invariably used in construction machines and vehicles in order to make the hydraulic systems compact. For instance, the brake circuit function within the dot-dash line limits shown in Fig. 3.76 can be interpreted as a mixed valve. The brake circuit is used to suppress the

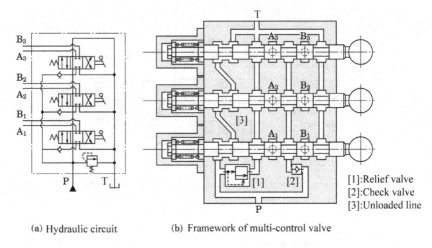

(a) Hydraulic circuit   (b) Framework of multi-control valve

[1]:Relief valve
[2]:Check valve
[3]:Unloaded line

Fig. 3.77   Typical multi control valve.

surge pressures that occur during the sudden stoppage of the actuator in the hydraulic circuits. Recently, the standardizations for the mixed valves with high universality have been developed.

Figure 3.77 illustrates a multi-control valve that may be classified as a mixed valve. The multi-control valve consists of three manual directional control valves, a relief valve and a check valve, and it controls the three actuators in a hydraulic power source. It is used for manual operation type directional control valve in hydraulic systems in construction machines.

3.3.6 *Modular Stack Valves*

Pipelines in hydraulic circuits give rise to vibration, noise and outer leakage, and moreover they make systems heavy and large. This also makes assembly work more difficult. Using modular stack valves such as sandwich valves, logic valves and cartridge valves, hydraulic systems can be configured without pipes connecting the hydraulic components.

3.3.6.1 *Sandwich Valve*

Stacking valves have fitting surfaces standardized by ISO 4401. Therefore it is possible to build hydraulic systems in which various sandwich valves may be interconnected in the stacking state by using through-bolts. Sandwich valves such as relief valves, directional control valves, pressure control valves and flow control valves are widely used in the hydraulic systems

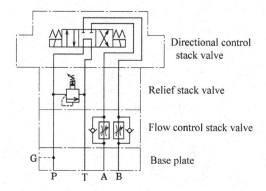

Fig. 3.78   Speed control circuit stacking sandwich valve.

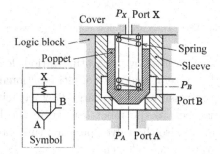

Fig. 3.79   Directional flow control logic valve element.

of machine tools, ships, rolling mills, etc. Figure 3.78 illustrates sandwich valves constituting a basic speed control circuit.

3.3.6.2 *Logic Valve*

(a) *Outline*

Logic valve elements are classified as directional flow control logic valve elements, pressure control logic valve elements and flow rate control logic valve elements. Figure 3.79 shows a directional flow control logic valve element consisting of a cartridge type poppet, a logic block, sleeve, a spring and a cover. A logic valve comprises several logic valve elements in a logic block with the standardized cover linking with each logic element port.

For instance, the directional control logic valve circuit shown in Fig. 3.81 acts as a four-way directional control valve operated by the piloted hydraulic pressure through a small solenoid directional control valve. That

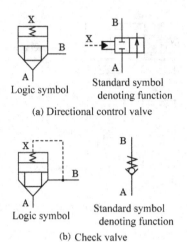

Logic symbol   Standard symbol
denoting function

(a) Directional control valve

Logic symbol   Standard symbol
denoting function

(b) Check valve

Fig. 3.80   Directional flow control logic valve.

is, the logic valve is constituted by integrating various logic valve elements so as to minimize the system, making it multi-functional and suitable for operation under high pressures and in large flow capacity hydraulic control systems. This is because the inner leakage in the logic valve is not only negligibly small but also the valve operation time can be regulated by controlling the flow rate in the pilot circuit.

(b) *Directional flow control logic valve element*

Figure 3.80(a) shows the hydraulic logic symbols denoting the function of the directional flow control logic valve element. The thrust force F acting on the logic poppet is defined as follows:

$$F = F_s + (S_A + S_B)P_X - S_A P_A - S_B P_B, \qquad (3.115)$$

where P_A is the pressure at port A, P_B is the pressure at port B, P_X is the pressure at pilot port X, S_A is the poppet area receiving pressure at the port A, S_B is the poppet area receiving pressure at the port B and F_s is the spring force.

The logic poppet opens when the force F shown in Eq. (3.115) takes a minus value. Therefore, the action of the poppet depends on following pressure conditions.

(1) The condition shutting the logic poppet is denoted as $P_X \geq P_A$ and $P_X \geq P_B$.

(2) The condition opening the logic poppet is denoted as $P_X = 0$ and $F_s < S_A P_A + S_B P_B$.

Linking the pilot port X to the port B as shown in Fig. 3.80(b), the thrust force F acting on the logic poppet is obtained by substituting $P_X = P_B$ into Eq. (3.115) as follows:

$$F = F_s + S_A(P_B - P_A). \tag{3.116}$$

Equation (3.116) governs the behavior of the poppet that opens under the condition $F_s + S_A(P_B - P_A) < 0$. Since the spring force F_s is negligibly small compared with the force $S_A(P_B - P_A)$ depending on the differential pressure, the logic flow control element linking port X to port B serves as a check valve.

Figure 3.81 illustrates a four-way directional flow control logic valve circuit operated by pilot pressures. The port A_P of the small solenoid operation type directional control valve is linked to the pilot ports X of the logic valve elements LV_1 and LV_2, whereas the solenoid type valve port B_P is linked to the pilot ports X of the logic valve elements LV_3 and LV_4. If the small solenoid type valve is operated to pilot the supply pressure to A_P, then the logic valve elements LV_1 and LV_2 close and the logic valve elements LV_3 and LV_4 are in free flow path states. Then, the working fluid flows into the left side cylinder chamber through the logic valve element

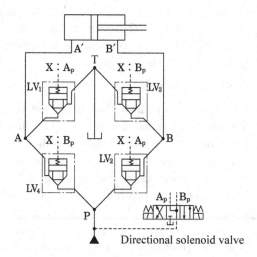

Fig. 3.81 Four-way directional control logic valve.

Table 3.10   Flow directions in the directional flow control logic valve system.

Operation condition	Flow direction	Moving direction of piston
Close:LV_1, LV_2 Open:LV_3, LV_4	P ➜ A ➜ A′ B′ ➜ B ➜ T	Right
Open:LV_1, LV_2 Close:LV_3, LV_4	P ➜ B ➜ B′ A′ ➜ A ➜ T	Left

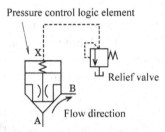

Fig. 3.82   Relief logic valve.

LV_4, and the working fluid in the right side cylinder chamber is discharged to the reservoir tank through the logic valve element LV_3.

When the small solenoid type valve is operated to pilot the supply pressure to port B_P, then the logic valve elements LV_3 and LV_4 close and the logic valve elements LV_1 and LV_2 are in free flow path states. Then, the working fluid flows into the right side cylinder chamber through the logic valve element LV_2, and the working fluid in the left side cylinder chamber is discharged to the reservoir tank through the logic valve elements LV_1. Table 3.10 shows the relationships between the operating conditions in logic valve elements and the flows in the circuit.

(c) *Pressure control logic valve element*

A pressure control logic valve element has the same configuration as a directional flow control valve except that it has a pilot line with a throttle opening the X-port chamber to A-port chamber, as shown in Fig. 3.82, and the pressure control spool element uses a spool instead of a poppet in the pressure control logic valve element, as shown in Fig. 3.83.

Recalling the balance piston type relief valve in Fig. 3.54, which has a pilot chamber opening to the pressure line through a throttle, the relief valve function is performed by regulating the pilot chamber pressure opening to the upper side of the main poppet valve. Therefore, as shown in Fig. 3.82, a large capacity logic relief valve is configured using both the pressure control

Pressure control
element using spool

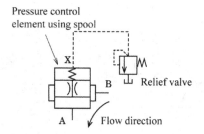

Fig. 3.83   Pressure reducing logic valve.

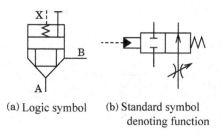

(a) Logic symbol   (b) Standard symbol
denoting function

Fig. 3.84   Flow rate control logic valve.

logic valve element and the small pilot relief valve regulating the pressure in the pilot port X.

Figure 3.83 illustrates a pressure reducing logic valve that uses both a pressure control spool element and a small relief valve. Regulating the pressure in the port X of the pressure control spool element using a small pilot relief valve, the system acts as a large capacity pressure reducing valve because it corresponds to the working principle of the pressure reducing valve shown in Fig. 3.55.

Sequence valves and counterbalances valves have similar configurations to the pressure reducing valves, although the pressure controlled in the system circuit is different. Therefore, they are able to be incorporated via a pressure control spool logic element.

(d) *Flow rate control logic valve element*

A flow rate control logic valve element has the same configuration as the directional flow control logic element, except that it provides a stopper to regulate the opening port area. It serves as a large capacity shut off valve with a variable throttle by means of small solenoid shut off valve and regulating the stopper position. Figure 3.84 shows a graphic symbol of a

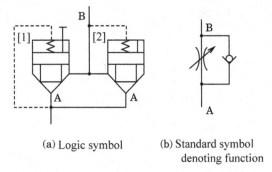

(a) Logic symbol   (b) Standard symbol
                              denoting function

Fig. 3.85   Logic check valve with variable throttle.

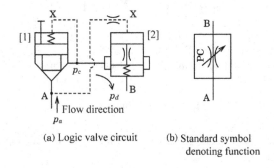

(a) Logic valve circuit   (b) Standard symbol
                                    denoting function

Fig. 3.86   Pressure compensated flow control logic valve.

flow rate control logic valve element and the graphic symbol denoting the function joining port X to port A.

Figure 3.85(a) shows a flow control logic valve system consisting of the flow rate control logic valve element [1] and the directional flow control logic valve element [2]. Though the flow from port A for port B passes through the logic valve element [2] freely, the flow from port B for port A passes only through the flow rate control logic valve element [1] regulating the flow rate. Figure 3.85(b) illustrates the function of the flow rate control logic valve system using standard symbols.

Figure 3.86 shows a pressure compensated flow control logic valve system. The flow control logic element [1] gives the logic valve port A, a preset opening area, and the movement of spool valve element [2] controls the variable spool port area, so as to keep the differential pressure $p_a - p_c$ constant, regardless of the changes of the pressure p_a and p_d.

(e) *Logic valve with multi-function*

Logic valve elements are not only applied in control valves such as pressure control valves, directional flow control valves and flow rate control valves, but also in multi-function logic valve systems provided with suitable pilot circuits. Figure 3.87 shows a multi-function logic valve system providing both a shut off function and a check valve function, through operating the pilot solenoid valve.

When the pilot solenoid valve is operated so that the spool takes the position [1] in Fig. 3.87(a), the logic element system serves as a check valve with free flow from the line-A to the line-B, because the port X opens to the line B through the shuttle valve. Operating the pilot solenoid valve so as to take spool position [2], the logic element blocks the flow between the line A and the line B because the line A traffics to the port X through a shuttle valve.

If the logic element in Fig. 3.87 is replaced by a flow rate control logic valve shown in Fig. 3.84, it becomes a multi-function valve acting both as a shut off valve and a flow rate control valve with a check valve in series by operating solenoid valve.

The logic valve circuit shown in Fig. 3.88(a) corresponds to the circuit in Fig. 3.81 except that the directional flow control logic valve elements LV_1, LV_2 and LV_4 in Fig. 3.81 are replaced by the counter balance logic valve circuit with $LV_1{}^*$, the flow rate control logic elements $LV_2{}^*$ and $LV_4{}^*$, respectively. Figure 3.88(b) illustrates the function of the logic valve circuit shown in Fig. 3.88(a) using standard symbols. When the pilot solenoid valve in Fig. 3.88(a) takes the spool position [1], the logic valve elements

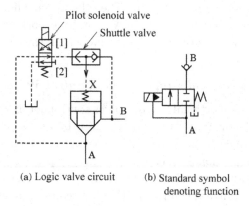

(a) Logic valve circuit   (b) Standard symbol
denoting function

Fig. 3.87   An example of multi-function logic valve.

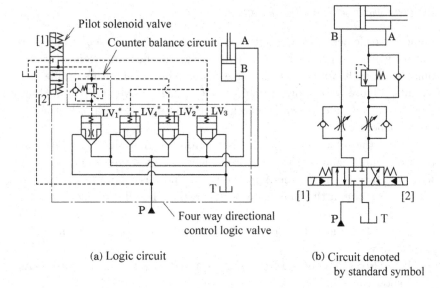

(a) Logic circuit

(b) Circuit denoted
by standard symbol

Fig. 3.88   Logic circuit with both functions of flow rate control and counter balance.

LV_3 and LV_4* shut the poppet ports, and the logic valve elements LV_1* and LV_2* open the poppet ports. Then, the working fluid flows into the B-side cylinder chamber through the flow rate control logic valve element LV_2*, and the fluid in the A-side cylinder chamber is exhausted to the reservoir tank through the logic valve element LV_1*.

When the solenoid valve takes the spool position [2], the logic valve elements LV_3 and LV_4* open the poppet ports and the logic valve elements LV_1* and LV_2* shut the poppet ports. Then the working fluid is supplied to the A-side cylinder chamber through LV_4* and the working fluid in B-side cylinder chamber is exhausted to the reservoir tank through LV_3.

3.3.7 *Servo Valves and Proportional Control Valves*

Servo valves and proportional control valves control not only the flow direction but also the flow rate. This is determined by the electric signal delivered from a controller in an electro hydraulic control system. Both the servo valve and the proportional control valves must be provided with a digital/analogue (D/A) and an analogue/digital (A/D) converter in order to join the controller to a digital computer. This is because they are operated on the basis of analogue electric signals.

3.3.7.1 *Servo Valves*

A servo valve is a four-way directional control valve continuously controlling the flow rate according to an analogue electrical signal. It is widely used in high-response and precise feedback control systems such as those found in aircraft, missiles and space shuttles. The servo valve may also be called an electro hydraulic servo valve.

The static accuracy of the servo valve is defined by the ratio of the electric output drift current ΔI to the rated electric current I_R in the amplifier driving the servo valve, and it is within the range from $\Delta I/I_R = 0.01$–0.03 in general. Figure 3.89 illustrates a typical two stage servo valve comprising a torque motor converting the electric current to the flapper deflections, a nozzle-flapper system converting the flapper deflections to a hydraulic pressure and a spool with a feedback spring, etc. When the electric current is not supplied to the torque motor coil, the spool is located at the central-reference position and the spool lands shut all the ports. Then, both pressure p_1 and p_2 in the nozzle chambers are identical because the flapper is located at the center between both nozzle injection ports. Supplying an electric current to the torque motor coil in accordance with an electric control voltage signal, the torque motor generates the torque to slightly incline the flapper so that the spool moves under the action of the differential pressure $p_1 - p_2$. If the flapper inclines slightly to the right, the pressure p_2 becomes larger than the pressure p_1. Consequently, the spool transfers to the left, and then it stops at the position where the spool force, which depends on the differential pressure $p_1 - p_2$, balances with the feedback spring force, since the flapper is connected with the spool by

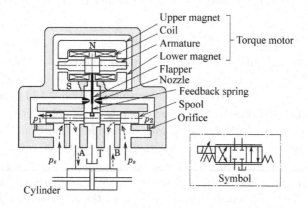

Fig. 3.89 Two stage type servo valve.

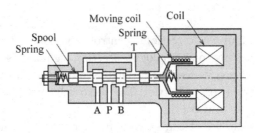

Fig. 3.90   Direct operated servo valve.

the feedback spring. Then, the flow rate into the A-side cylinder chamber through the servo valve is proportional to the electric current, because the opening areas of the spool lands are proportional to the feedback spring force in relation to the electric current. Therefore, the servo valve is able to control the actuator speed and the moving direction of the actuator by supplying the electric current to the coil of the servo valve.

Figure 3.90 illustrates a direct operated servo valve utilizing a moving coil to drive a spool. The moving coil is placed at the clearance in the magnetic field as shown in Fig. 3.90. Supplying electric current to the moving coil crossing the magnetic field, the magnetic force that moves the spool axially is generated in accordance with the Fleming's left hand law. Therefore, the spool transfers to a position where the magnetic force balances with the force of the spring connecting to the spool so that the flow rate through the servo valve is in proportion to the electric current level.

The response characteristics of a direct operated servo valve is about 200 Hz at the phase angle $-90°$, although the flow rate capacity is less than 100 L/min. It is widely used as a small/middle capacity servo valve.

3.3.7.2 *Proportional Solenoid Control Valve*

Generally, solenoid magnetic force is proportional to the square of the coil current and inversely proportional to the square of the clearance between the plunger and the electric magnet as shown in Eq. (3.114) previously. However, a proportional control valve controls the spool positions using an improved solenoid in which the magnetic force is proportional to the coil current regardless of the plunger position. A proportional solenoid control valve is more resistant to fluid contaminants than a servo valve is, but, on the other hand, the static and the dynamic characteristics are inferior to those of the servo valves. They are used as the control valve in various machines such as injection molding machines, press machines, load test

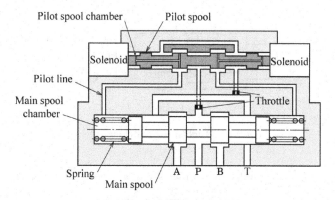

Fig. 3.91 Two stage proportional solenoid directional control valve.

machines, conveying installations, etc. A proportional solenoid control valve has the same functions as a servo valve though the response time is nearly ten times larger than that of a servo valve.

Figure 3.91 illustrates a two stage proportional solenoid valve. If the pilot spool transfers slightly to the left by giving an electric current to the proportional solenoid control valve, then the pressure p_p is produced in both the pilot left spool chamber and the main left spool chamber due to the flow passing through pilot lines. Then, the pilot spool stops at the position where the pilot spool force, which depends on the pressure p_p, balances with the solenoid magnetic force, and concurrently the main spool transfers to the position at which the main spool force , which depends on the pressure p_p, balances with the main spring force. Consequently, the main spool movement makes the opening port-area proportional to the electric current level. The response time for a full stroke of the main spool is about 50 ms for a proportional control valve with a rated flow rate in the region of 60 L/min.

Using the proportional solenoid control valve, various control valves can be configured and used in practical applications, including such relief valves, pressure reducing valves and pressure compensated flow control valves.

3.3.8 *High-speed Switching Valves*

A high-speed switching valve, such as that shown in Fig. 3.92 is a high response shut off valve operated by on/off signals, and the switching time is within the range of 2.0 to 3.0 ms. The maximum valve lift is within 0.3 mm. A typical performance of the high-speed switching valve is shown

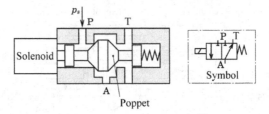

Fig. 3.92　High-speed switching valve.

Table 3.11　High-speed switching valve performance.

Rated pressure	17.5 MPa
Rated flow rate	8 L/min
Limited frequency	50 Hz
ON switching time	4 to 6 ms
OFF switching time	1.5 to 2.5 ms
Solenoid voltage	12 V
Solenoid current	2.6 A

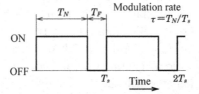

Fig. 3.93　PWM signal.

in Table 3.11. Since the rated flow rate of a high-speed switching valve is in general less than 12 L/min, it may need to be amplified using a logic valve element when a larger flow rate is required. Although the high-speed switching valves shut or open the port on the basis of on/off signals, they can be also controlled by using pulse width modulation (PWM) signals transformed from the analogue signals.

Figure 3.93 illustrates PWM signals comprising the constant frequency pulse width signals with the operating time T_N in according with the analogue signal. The modulation rate τ in PWM is defined as follows:

$$\tau = \frac{T_N}{T_s} = \frac{T_N}{T_F + T_N}, \qquad (3.117)$$

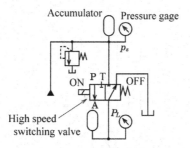

Fig. 3.94 PWM pressure circuit.

where T_s is the time cycle period in PWM, T_N is the operating period in PWM and T_F is the un-operating period in PWM.

Figure 3.94 illustrates a pressure control circuit utilizing a high-speed control switching valve operated by PWM signals. The high-speed switching valve supplies the working fluid into the A-port line circuit for the operating period T_N in PWM signals, because the port P opens to port A for the time period T_N, and then the flow rate Q_1 is described as follows:

$$Q_1 = \tau C_d A_p \sqrt{\frac{2(p_s - p_L)}{\rho}}, \qquad (3.118)$$

where $\tau = T_N/T_s$ and A_p denote the modulation rate and the opening area of the port A, respectively, p_s is the pressure in the supply line and p_L is the mean pressure in the A-port line. The working fluid in the A-port line circuit is discharged to the reservoir tank in the non-operation period T_F in PWM signals because port A opens to port T, and then the flow rate Q_2 is described as follows:

$$Q_2 = (1 - \tau) C_d A_p \sqrt{\frac{2p_L}{\rho}}. \qquad (3.119)$$

Recalling Eqs. (3.118) and (3.119) and $Q_1 = Q_2$, the pressure p_L in the steady state is derived as follows:

$$p_L = \frac{p_s}{\left(\frac{1-\tau}{\tau}\right)^2 + 1}. \qquad (3.120)$$

Figure 3.95 shows the relationships between the modulation rate τ and the dimensionless mean pressures p_L/p_s derived from Eq. (3.120).

Figure 3.96 illustrates a servo system using high-speed switching valves. The electric voltage K_e amplifying the error signal e is converted into a

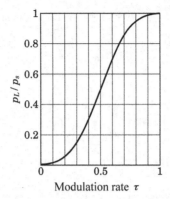

Fig. 3.95　Dimensionless output pressure in PWM pressure circuit.

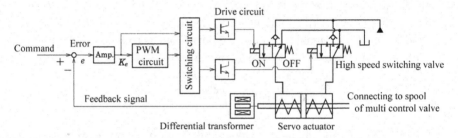

Fig. 3.96　Servo system using high-speed switching valves.

PWM signal, sent to the switching circuit to select the drive circuit depending on the plus or minus sign of the voltage K_e. Then, the selected drive circuit operates the high-speed switching valves in accordance with the PWM signals so that the servo-actuator piston moves to decrease the control error signal e. Therefore, the position of the servo actuator piston is controlled to match the command value. Giving the off-condition for both high-speed switching valves, the servo actuator piston is returned to the central position by the spring force.

3.4 Hydraulic System Components

A hydraulic system is comprised of not only hydraulic pumps, hydraulic actuators and various control valves, but also the miscellaneous components called hydraulic system components such as reservoirs, accumulators, oil filters, heat exchangers, transmission lines and fittings.

3.4.1 *Accumulators*

3.4.1.1 *Outline*

An accumulator absorbs surplus fluid energy during the operation of hydraulic system and uses it as a supplemental power source when the system temporarily requires power that exceeds the pump capacity. Accumulators also have the following uses:

- Providing power in a hydraulic system with a short driving time;
- Providing emergency power in the case of power failure;
- Suppressing surge pressures in hydraulic systems;
- Eliminating pressure ripples;
- Compensating for fluid leakage and changes in temperature.

Accumulators may be mainly classified spring-loaded or gas-loaded. Figure 3.97 shows a spring-loaded accumulator that stores fluid pressure energy through a preloaded spring held by the hydraulic system source. The preloaded spring pushes the stored working fluid out into the system when the system requires supplemental fluid power. The spring-loaded accumulator is used at low pressure as an accumulator with small capacity, but it can be used in a wide range of temperatures.

Gas-loaded accumulators store the fluid pressure energy against the compressed gas in accordance with the Boyle's law. The bladder accumulator shown in Fig. 3.98 is the most typical gas-loaded accumulator, comprising an enclosed shell, an elastic bladder confined with preloaded compressed gas, a gas-charging valve and an oil port.

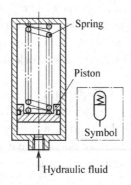

Fig. 3.97 Spring-loaded accumulator.

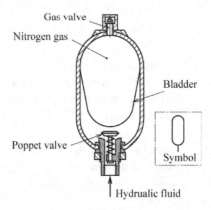

Fig. 3.98   Bladder accumulator.

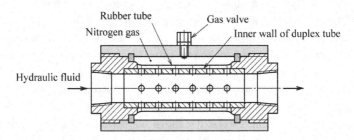

Fig. 3.99   In-line accumulator.

The in-line accumulator shown in Fig. 3.99 is a gas-loaded type comprising an elastic bladder inserted between the inner and the outer cylindrical wall of a duplex pipe. The inner cylindrical wall has many holes to ensure that the rubber surface comes in contact with the working fluid in the duplex pipe, and the compressed gas is confined in the elastic bladder. The in-line accumulator is used to suppress shock pressures and to eliminate pressure ripples in pipelines.

As shown in Fig. 3.100, a diaphragm accumulator comprises a spherical shell inside which is an elastic diaphragm that separates the interior into a compressed gas chamber and a working fluid chamber. Although diaphragm accumulators generally do not have a large capacity, they are often used in aircraft hydraulic systems because they have a high power density.

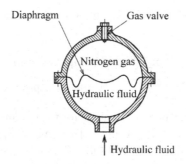

Fig. 3.100   Diaphragm accumulator.

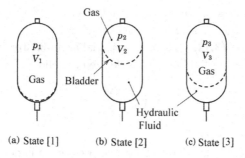

(a) State [1]   (b) State [2]   (c) State [3]

Fig. 3.101   Working states of accumulator.

3.4.1.2 *Accumulator Capacity to Provide Supplemental Fluid Energy*

Figure 3.101 illustrates the working states of a bladder accumulator, where V_1, V_2 and V_3 denote the confined gas volume at the absolute pressure p_1, p_2 and p_3 in states [1], [2] and [3], respectively. As shown in Fig. 3.101(b), it is assumed that the confined gas volume V_1 at pressure p_1 in the initial state [1] changes to volume V_2 at pressure p_2 by absorbing the fluid energy in the hydraulic system into the compressed gas in the accumulator. When the supplemental fluid power is required, the gas volume V_2 at pressure p_2 changes to the volume V_3 at pressure p_3, releasing the stored fluid energy into the system, as shown in Fig. 3.101(c).

The working fluid volume ΔV discharged from the accumulator into the system is given as follows:

$$\Delta V = V_3 - V_2. \tag{3.121}$$

The equation of state for the gas confined in the bladder are expressed as follows:

$$p_1 V_1^{n_c} = p_2 V_2^{n_c}, \quad \text{(Compression process)} \tag{3.122}$$

$$p_2 V_2^{n} = P_3 V_3^{n}, \quad \text{(Expansion process)} \tag{3.123}$$

where n and n_c are polytropic exponents for the expansion and the compression process, respectively. The value of the exponents is $n_c = n = 1$ for the isothermal change, and $n_c = n = 1.4$ for adiabatic change in diatomic gases (e.g. N_2), and it takes $n_c = n = 1.66$ for the adiabatic change in monatomic gases (e.g. He). Typically, the gas confined in the bladder is nitrogen. When the gas volume confined in the bladder compresses or expands as slowly as possible, the process becomes isothermal. Conversely, the process becomes adiabatic when the gas volume changes quickly.

Recalling Eqs. (3.121)–(3.123), an initial accumulator gas volume V_1 to provide supplemental fluid energy is derived as follows:

$$V_1 = \frac{\Delta V}{\left(\frac{p_1}{p_2}\right)^{1/n_c} \left[\left(\frac{p_2}{p_3}\right)^{1/n} - 1 \right]}. \tag{3.124}$$

Since the energy of gas volume V at the pressure p is denoted as pV, the efficiency η of the accumulator is defined as follows:

$$\eta = \frac{p_3 V_3}{p_1 V_1}. \tag{3.125}$$

Recalling Eqs. (3.122), (3.123) and (3.125), the flowing equation is derived.

$$\left(\frac{p_1}{p_2}\right)^{1/n_c} = \eta \frac{p_1}{p_3} \left(\frac{p_3}{p_2}\right)^{1/n}. \tag{3.126}$$

Substituting Eq. (3.126) into Eq. (3.124) gives

$$V_1 = \frac{\Delta V \left(\frac{p_3}{p_1}\right) \left(\frac{p_2}{p_3}\right)^{1/n}}{\eta \left[\left(\frac{p_2}{p_3}\right)^{1/n} - 1 \right]}. \tag{3.127}$$

Applying Eq. (3.127), the empirical conditions of $0.9p_2 \geq p_1 \geq 0.25p_3$ and accumulator efficiency $\eta \approx 0.95$ are adopted. Then, polytrophic exponents n_c and n for nitrogen gas are taken to be the values shown in Fig. 3.102.

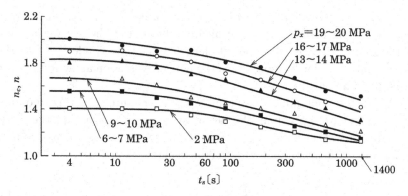

Fig. 3.102 Polytropic exponent of nitrogen gas.

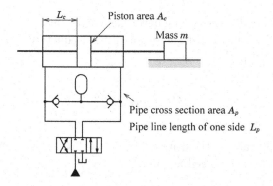

Fig. 3.103 Hydraulic system with an accumulator to absorb surge pressure.

3.4.1.3 *Accumulator Capacity to Absorb Surge Pressure*

When an actuator that is under load is suddenly stopped by shutting the valve ports, a surge pressure arises in the hydraulic system. This is because the kinematic energy of the actuator is absorbed into the fluid pressure energy by producing the slight volume change of the quasi-incompressible working fluid. The surge pressure is unwanted, and an accumulator may be used to absorb it.

Let us consider the accumulator's capacity to absorb surge pressure. Figure 3.103 shows a hydraulic system with the accumulator to absorb the surge pressure corresponding to the sudden stoppage of the actuator loaded with a mass m. When the actuator is moving at the velocity v_c, the kinematic energy W of the mass is expressed as follows:

$$W = \frac{v_c^2}{2} \left[m + \rho L_c A_c \left(1 + \frac{L_p A_p}{L_c A_c} \right) \right], \tag{3.128}$$

where A_c is the effective piston area receiving pressure, L_c is the half stroke of piston being assumed that the piston is located at the central position of the cylinder, L_p is the pipe line length from the cylinder port to the directional control valve port, A_p is the pipe cross-section area, and ρ is the working fluid density.

When the accumulator gas volume V_n at pressure p_n acting upon the load changes to the volume V_m at pressure p_m by absorbing the surge pressure due to the sudden blocking of the piston, the energy equation may be formulated using the infinitesimal gas volume change dV

$$\int\limits_{v_n}^{v_m} p\,dV + W = 0. \tag{3.129}$$

Now, if V_m stands for the accumulator gas volume at the acceptable maximum system pressure p_m, then substituting a equation of state $pV^{n_c} = C = \text{const.}$ into Eq. (3.129) gives

$$\int\limits_{v_n}^{v_m} p\,dV = \int\limits_{v_n}^{v_m} CV^{-n_c}\,dV = \frac{p_n V_n - p_m V_m}{n_c - 1}. \tag{3.130}$$

The equation of state for the confined gas in the process whereby the kinetic energy is absorbed is defined as follows:

$$p_n V_n^{n_c} = p_m V_m^{n_c} = p_1 V_1^{n_c}. \tag{3.131}$$

Recalling Eqs. (3.128)–(3.131), the gas volume V_n at the normal pressure p_n acting upon the load is denoted as follows:

$$V_n = \frac{v_c^2\left[m + \rho L_c A_c\left(1 + \frac{L_p A_p}{L_c A_c}\right)\right](n_c - 1)}{2p_n\left[\left(\frac{p_m}{p_n}\right)^{(n_c-1)/n_c} - 1\right]}. \tag{3.132}$$

The initial accumulator gas capacity V_1 at pressure p_1 is derived from Eqs. (3.131) and (3.132) as follows:

$$V_1 = V_n\left(\frac{p_m}{p_1}\right)^{1/n_c} = \frac{v_c^2\left[m + \rho L_c A_c\left(1 + \frac{L_p A_p}{L_c A_c}\right)\right](n_c - 1)\left(\frac{p_n}{p_1}\right)^{1/n_c}}{2p_n\left[\left(\frac{p_m}{p_n}\right)^{(n_c-1)/n_c} - 1\right]}.$$

$$\tag{3.133}$$

The pressure ratio p_n/p_1 usually takes a value within the range 0.8 to 0.9 when the accumulator is absorbing surge pressure or fluctuations of pump output pressure.

3.4.2 *Heat Exchangers*

When the hydraulic system is in operation, part of the fluid power is converted into heat energy due to the fluid friction involved in the flow passing through the fittings, throttles and clearances in the system. The converted heat energy per unit time is the difference between the effective output of the hydraulic system and the fluid power input by an ideal pump. Then, the working fluid temperature in the system rises until the converted heat energy per unit time balances the heat energy per unit time dissipating to the atmosphere or that of the cooling medium in the heat exchanger.

In operational systems the working fluid temperature should be maintained below 60°C because the reduction of the fluid viscosity increases internal leakage, and also gives rise to a range of problems such as the rapid wear of components, and seizures in the movable parts. If the fluid temperature exceeds that limit for a long time, the working fluid will be oxidized giving rise to sludge and, in consequence, the life of the hydraulic system will be significantly reduced. Therefore, a heat exchanger should be installed so as to maintain the fluid temperature within those limits.

Heat exchangers are divided into the water-cooled type and the air-cooled type. Figure 3.104 illustrates a water-cooled heat exchanger. The working fluid flows over the cooling pipes in which the cooling water is provided, and it is directed perpendicular to the cooling pipe axis by means of the baffle plates. In air-cooled heat exchangers, a rotating fan generates the flow of natural air over a radiator made of finned tubes in which the

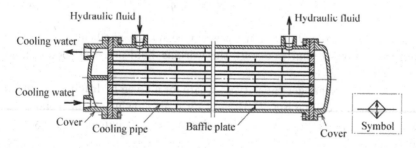

Fig. 3.104   Water cooled heat exchanger.

working fluid is cooled. Though the water-cooled type is a universal device, the air-cooled type is preferred because it is inexpensive and non-corrosive, and moreover it can be usefully applied in both mobile and stationary hydraulic systems, and used in places where there is no source of cooling water.

Using the heat transfer coefficient K_h indicating the heat exchanger performance given in the manufacturer's specification, the heat quantity H per unit time dissipated by the heat exchanger is defined as follows:

$$H = K_h A \Delta\theta, \qquad (3.134)$$

where A is the heat transfer area and $\Delta\theta$ is the mean differential temperature between working fluid and the cooling medium in the heat exchanger. Though the heat transfer coefficient takes the different value depending on the heat exchanger type and its construction, it typically falls in the range between $175\,\mathrm{W}/(^\circ\mathrm{C m^2})$ and $350\,\mathrm{W}/(^\circ\mathrm{C m^2})$.

The amount of water in a water-cooled heat exchanger generally ranges between 50% and 100% of the cooled hydraulic fluid. When the logarithmic form of the mean temperature difference is adopted in Eq. (3.134), the better estimate may be provided. The logarithmic mean temperature $\Delta\theta$ is defined as follows:

$$\Delta\theta = \frac{(T_1 - t_2) - (T_2 - t_1)}{\ln \frac{T_1 - t_2}{T_2 - t_1}}, \qquad (3.135)$$

where T and t denote the temperature of the working fluid and the cooling water, respectively, and the suffixes 1 and 2 denote the locations at the inlet and at the outlet of the heat exchanger, respectively. The heat exchange equation in the heat exchanger is described as follows:

$$H = \rho_o q_o c_o (T_1 - T_2) = \rho_w q_w c_w (t_2 - t_1), \qquad (3.136)$$

where ρ, q and c denote the density, flow rate and specific heat, respectively, and the suffixes o and w denote the working fluid and the cooling fluid, respectively.

The heat exchanger capacity adequate for a given hydraulic system can be calculated recalling Eqs. (3.134)–(3.136). A following example shows the procedure applied to calculate the suitable heat exchanger capacity for a hydraulic system.

[Example 3.2]

The heat quantity per unit time $H = 6.05\,\mathrm{kW}$ is exchanged by a heat exchanger when cooling water with an entrance temperature $t_1 = 25°\mathrm{C}$ flows into the heat exchanger at the flow rate $Q = 24$ L/min and the working fluid with an entrance temperature $T_1 = 50°\mathrm{C}$ flows through it at the flow rate $Q = 60$ L/min.

(1) Calculate the value of $K_h A$ denoting the product of the heat transfer coefficient K_h by the heat transfer area A of the heat exchanger.
(2) Discuss whether the heat exchanger is able to lower the temperature of working fluid to $T_2 = 50°\mathrm{C}$ provided the cooling water with the entrance temperature $t_1 = 22°\mathrm{C}$ flows into it at the rate $Q = 40$ L/min and the working fluid with the entrance temperature $T_1 = 55°\mathrm{C}$ flows through it at the rate $Q = 60$ L/min.

Where hydraulic fluid density ρ_o, specific heat c_o of the working fluid and specific heat c_w of the cooling water are $\rho_o = 870\,\mathrm{kg/m^3}$, $c_o = 1.88\,\mathrm{kJ/(kg°C)}$ and $c_w = 4.19\,\mathrm{kJ/(kg°C)}$, respectively.

[Solution 3.2]

(1) The exit temperature T_2 of the working fluid and the exit temperature t_2 of the cooling water are obtained from Eq. (3.136) as follows:

$$T_2 = T_1 - \frac{H}{\rho_o q_o c_o} = 50 - \frac{6050}{870 \times \left(\frac{60 \times 10^{-3}}{60}\right) \times (1.88 \times 10^3)} = 46.3°\mathrm{C},$$

$$t_2 = t_1 + \frac{H}{\rho_w q_w c_w} = 25 + \frac{6050}{1000 \times \left(\frac{24 \times 10^{-3}}{60}\right) \times (4.19 \times 10^3)} = 28.6°\mathrm{C}.$$

Substituting the values of these temperatures into Eq. (3.135) gives

$$\Delta\theta = \frac{(50 - 28.6) - (46.3 - 25)}{\ln\left(\frac{50 - 28.6}{46.3 - 25}\right)} = 21.4°\mathrm{C}.$$

Substituting $H = 6050\,\mathrm{W}$ and $\Delta\theta = 21.4°\mathrm{C}$ into Eq. (3.134) gives

$$K_h A = \frac{6050}{21.4} = 283\,\mathrm{W/°C}. \tag{1}$$

(2) The exchanged heat quantity H is calculated recalling Eq. (3.136):

$$H = 870 \times \left(\frac{60 \times 10^{-3}}{60} \right) \times 1.88 \times 10^3 \times (55 - 50) = 8180 \, \text{W}. \quad (2)$$

Therefore, the cooling water temperature t_2 at the exit is derived from Eq. (3.136)

$$t_2 = t_1 + \frac{H}{\rho_w q_w c_w} = 22 + \frac{8180}{1000 \times \left(\frac{40 \times 10^{-3}}{60} \right) \times (4.19 \times 10^3)} = 24.9°\text{C}.$$

Logarithmic temperature $\Delta\theta$ is obtained from Eq. (3.135)

$$\Delta\theta = \frac{(55 - 24.9) - (50 - 22)}{\ln \left(\frac{55 - 24.9}{50 - 22} \right)} = 29°\text{C}.$$

Since the heat exchanger performance $K_h A = 283 \, \text{W}/°\text{C}$ is implied by Eq. (1), the exchangeable heat quantity H_x is calculated from Eq. (3.134):

$$H_x = 283 \times 29 = 8210 \, \text{W} = 8.21 \, \text{kW}.$$

The heat exchanger should be able to cool the working fluid to the temperature 50°C at the exit tentatively, because the exchangeable heat quantity $H_x = 8210 \, \text{W}$ is larger than the exchanged heat quantity $H = 8180 \, \text{W}$ shown in Eq. (2) though the margin is very small.

3.4.3 *Filters*

It is essential for the reliability of hydraulic control systems that the working fluid has the required performance parameters. However, the fluid tends to be contaminated by the following factors:

- Wear products generated in the components;
- Chemical reaction products such as sludge;
- Contaminants entering through the maintenance practices.

The presence of contaminants leads to a gradual deterioration of working fluid performance and it may cause catastrophic failures of the hydraulic systems. Filters are devices designed to arrest the contaminants in the working fluid by making the working fluid pass through a porous element made of filtering papers, coiling wires, sintered metals, etc.

A filter may be rated by the nominal size of the particles removed by the filter. However, the most reliable method to define filtration performance is to use the beta ratio defined in ISO 4572 as follows:

$$\beta_x = \frac{n_1}{n_2}, \qquad (3.137)$$

where n_1 denotes the number of particles over $x\,\mu\text{m}$ included in a constant fluid volume before it flows into the filter, n_2 denotes the numbers of particles over $x\,\mu\text{m}$ in the volume after it has passed through the filter. For instance, $\beta_{12} = 100$ indicates that the particles over $x = 12\,\mu\text{m}$ in the fluid is removed over 99% by the filter.

Suction filters consist of a porous medium supported with a metal frame whose end has a flange connecting to the suction pipe. They are installed in the suction line to trap large contaminants damaging the pump. Generally the removal of large contaminants is achieved by using suction filters with a nominal filtration rating of $44\,\mu\text{m}$, $74\,\mu\text{m}$ or $105\,\mu\text{m}$. In contrast, line filters with a nominal filtration rating of $1, 3, 6, 12, 25$ or $40\,\mu\text{m}$ are used in pressure lines, in order to remove fine contaminants and thus protect those system components that are more sensitive to dirt than pumps.

In modern hydraulic systems it is important to arrest contaminant particles smaller than $5\,\mu\text{m}$ by means of filters, because otherwise such particles may enter into the gap between moving/sliding surfaces and abrade the surface. The abraded particles thus produced then become abrasives themselves, adding to the problem.

Figure 3.105 illustrates a line filter into which working fluid flows through the inlet port. The working fluid passes through the cylindrical

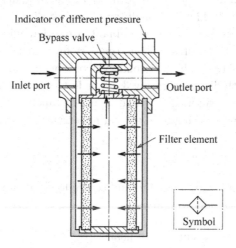

Fig. 3.105   Line filter.

[1] Off-loop filter
[2] Suction filter
[3] Pressure line filter
[4] Drain filter
[5] Servo valve filter
[6] Return line filter
[7] Bleed-off filter

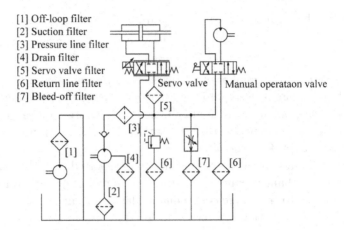

Fig. 3.106   Examples of filters installed in hydraulic systems.

porous medium from the outer periphery side, and then flows out of the outlet port. When the differential pressure between the inlet and the outlet of the filter becomes large due to the porous medium being plugged with contaminants, a part of working fluid flows through the bypass valve opened by the differential pressure. At the same time, the differential pressure produces a signal that the porous medium needs to be replaced.

Figure 3.106 illustrates some examples of a hydraulic system provided with various filters. The off-loop filter system [1] is used to promote further cleaning of the fluid or may be used in hydraulic systems when line filters cannot be installed. The suction filter [2] is installed at the suction line of the pump to prevent large contaminants from damaging the pump, and the line filter [3] in the pressure line removes fine particles before they get diffused in the system. The dirt sensitive components such as servo valves require filters with filtration rating $3\,\mu$m or less to be installed as shown in filter [5]. The return filters [4], [6] and [7] are installed in case contaminants from the returning working fluid should penetrate into the system.

3.4.4 *Reservoirs*

The reservoir is not only the main fluid container in a hydraulic system but it also serves as the following uses:

- Device to cool the accumulated heat;
- Settling tank to separate the contaminants and air bubbles from the fluid;

- Space to accommodate the expansion of the fluid due to the temperature changes;
- Supporting base for the pump-motor assembly containing the hydraulic circuit.

The reservoir is usually a rectangular, closed container made of mild steel plate with the thickness of 3 mm or more and it provides a sufficiently large surface area for adequate dissipation of accumulated heat in working fluid. Therefore, the reservoir has generally three up to five times as large as the maximum pump displacement per minute.

Figure 3.107 shows a typical non-pressurized closed type reservoir tank. The top plate of the reservoir should have a sufficiently rigid structure to endure shocks due to vibration and the mechanical misalignment of the pump/motor unit mounted on it. A port serving as both the oil filler and the breather filtering air is set up on the top plate of the reservoir. The inside of the reservoir is divided into the sink side and the return side by a baffle plate with holes, in order to settle the particles at the bottom and to let out air from the fluid. The bottom plate of the reservoir has a slope in the range from 1/20 to 1/25 for easy removal of the old working fluid for maintenance purposes, and it is raised at least 150 mm from the ground by mounting it on the legs. This is to ensure better air circulation. Side plates are provided in order to facilitate inspection and cleaning. The end of the return pipe is usually cut at an angle of 45° and it is directed towards the reservoir wall so as not to disturb the contaminants deposited at the bottom. Other accessories may be fitted. These may include fluid level indicators, fluid temperature indicators, magnetic trappers to trap ferrous particles in the fluid, etc.

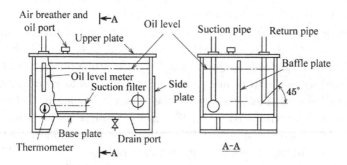

Fig. 3.107   Typical closed type reservoir tank.

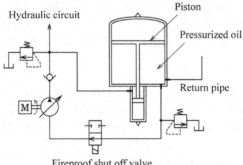

Fig. 3.108   Pressurized reservoir.

Figure 3.108 illustrates a pressurized reservoir in an aviation hydraulic system. The working fluid in the reservoir is controlled to hold the pressure at about 0.3 MPa in order to prevent cavitations at the pump suction port because it is used at low atmospheric pressure.

Special hydraulic source units fixing the pump/motor assembly within the reservoir with the working fluid have been developed for purpose of noise protection.

3.4.5 *Hydraulic Pipes and Fittings*

3.4.5.1 *Pipes*

Seamless cold drawn steel tubes are the most common choice for hydraulic pipes because they have a very smooth inner wall and they may be manufactured to a high degree of accuracy for any wall thickness. However, hydraulic pipes made of the aluminum/titanium alloy are used in aviation engineering to make the hydraulic systems lighter.

Although pipe size is generally specified by inner diameter, precision drawn steel tube is denoted by the nominal bore size. This is not always the actual inner tube diameter, because the steel tube outer diameter size is standardized so as to retain the same outside threads regardless of the change in the wall thickness of the tube. There are some inched nominal bore size tube-standards based on ISO specification, e.g. the British standard BS4368, Japanese standard JIS G3454, American standard SAE J154.

As shown in Fig. 3.109, the tube-bending radius R is in general limited as $R/d_i \geq 2.5$ because the stress between the outer and inner wall in the pipe increases considerably as the bending curvature becomes smaller.

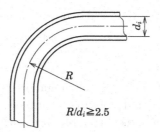

Fig. 3.109 Curvature in piping.

Table 3.12 Recommended fluid velocity in tubes.

Suction pipe		Pressure pipe ($\nu = 30$–$150\,\mathrm{mm^2/s}$)		Return pipe
Kinematic viscosity ν [mm²/s]	Mean velocity u [m/s]	Pressure p [MPa]	Mean velocity u [m/s]	Mean velocity u [m/s]
150	0.6	2.5	2.5–3	1.7–4.5
100	0.75	5.0	3.5–4	
50	1.2	10	4.5–5	
30	1.3	20	5–6	

Hydraulic hose is a flexible tube composed of a rubber inner reinforced with fiber or wire braiding, and an outer cover that protects the inner tube. Standardized fittings are set on both the ends of the hose. It is used for the piping in systems in which components may need to be moved with respect to other components. Hydraulic hose also has the benefit of absorbing shock and vibration. Hydraulic hoses in accordance with the ISO specification are now commercially available.

Tube sizes in a hydraulic system are selected basing on the following procedures (1) to (4).

(1) Table 3.12 shows the recommended mean velocity u of the fluid in tubes for various conditions. Since the flow rate through the tube is given by $Q = (\pi d_i^2/4)u$, the inner tube diameter d_i may be described as follows:

$$d_i = \sqrt{\frac{4Q}{\pi u}}. \tag{3.138}$$

Using a mean velocity $u = u_1$ recommended in Table 3.12, a preferable value of the inner diameter d_i may be temporarily decided for the flow rate Q in the hydraulic system.

(2) A tube wall thickness δ is calculated using the Barlow's formula

$$\delta = \frac{p_m d_i}{2\left(\frac{\sigma_t}{S} - p_m\right)}, \tag{3.139}$$

where p_m, σ_t and S denote the maximum pressure, the tensile strength of the pipe materials and the safety factor, respectively. The relevant safety factors are adopted in accordance with the conditions in the hydraulic system as follows:

- $S = 4$: Pressure in the pipe does not exceed 17 MPa,
- $S = 6 - 8$: Surge pressures occur in the circuit, and mechanical stresses act on the tubes though the rated maximum pressure in the circuit does not exceed 17 MPa,
- $S = 10$: Sever surge pressures occur in the circuit, and considerable mechanical stresses act on the tubes.

(3) Select the adequate standardized tube according to the inner diameter and wall thickness calculated by the above procedures (1) and (2).
(4) Calculate the pressure loss in the pipeline for the selected tube recalling Eqs. (2.75)–(2.78) in Section 2.3.3. If the pressure loss in pipeline is too large, reselect the suitable tube diameter and the wall thickness according to the procedures from (1) to (3) on the conditions that the mean velocity u is altered, and then reselect the preferable standardized tube.

3.4.5.2 *Pipe Fittings*

As shown in Fig. 3.110, the main types of pipe fittings to consider are screwed joints, welded joints, flanged coupling joints and compression joints. Figure 3.110(a) shows a T-connector with tapered thread ports which joins three pipes. It has a straight thread at the outer periphery. A screw joint that uses a straight and a tapered thread seals the fluid by the interacting face in a screw joint. The typical tapered thread for screwed joints is the dry seal American Standard pipe taper threads (NPTF). Figure 3.110(b) shows an elbow welded to pipes. It is generally advisable to avoid welding joints if possible, because of the risk of producing scales in the inside surface of the pipe. Figure 3.110(c) illustrates the one sided configuration of the flange coupling. In a flanged coupling, the fluid is sealed by compressing the O-ring, which is in a circular ditch on the flange surface, against the other side. Figure 3.110(d) illustrates a flareless type of compression union joint connecting two tubes. Flareless compression fittings seal by compressing the

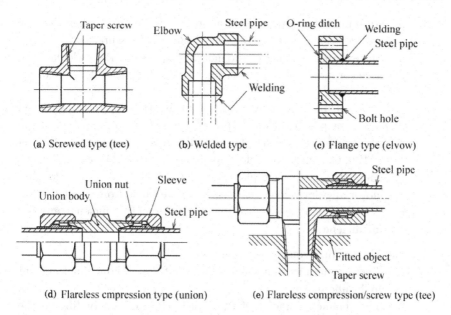

(a) Screwed type (tee) (b) Welded type (c) Flange type (elvow)

(d) Flareless cmpression type (union) (e) Flareless compression/screw type (tee)

Fig. 3.110 Fittings.

metal surfaces between the surfaces of a ferrule fitted to the pipe and the connector element. In contrast, flared compression fittings seal using metal surfaces between the flared pipe and the connector element. Figure 3.110(e) illustrates a flareless compression/screw joint type T-connector connected to the hydraulic component by a screw joint.

Hydraulic components are typically designed to be handled by any pipe fitting method such as: directly jointing the pipe to the hydraulic component port by using a screw joint; jointing pipe connectors to the hydraulic component using a screw joint; jointing a flange type connector with the welded pipe to the hydraulic component.

Problems

3.1 A pump rotating at $N = 1800$ min^{-1} displaces hydraulic fluid at the rate $Q = 48$ L/min. The pump outlet pressure is $p_s = 28$ MPa. Calculate the values of (1) to (3) when the volumetric efficiency of the pump is 0.9 and the torque efficiency of the pump is 0.87.

(1) Swept volume per unit rotation $2\pi D_p$ of pump.

(2) Power P required to drive the pump.

(3) Torque T_a driving the pump.

3.2 A hydraulic motor rotates the load at $N = 1500$ min^{-1} when the flow rate of $Q = 36$ L/min is supplied to it at a pressure $p_s = 14$ MPa and the exhaust pressure of the motor is $p_0 = 0.2$ MPa. Calculate the effective torque T_a of the motor for an overall motor efficiency 0.8.

3.3 A variable displacement hydraulic motor is driven by rotating a fixed displacement pump with a swept volume of $D_p = 5.2$ cm^3/rad at the rotational speed of $N = 1200$ min^{-1}. The pump pressure p_s is limited to 10 MPa. Calculate the following quantities from (1) to (3) when the maximum inclined swash plate angle of the motor is $\alpha_m = 20°$ and the motor angular velocity ω must be in the range from $\omega = 6.6$ rad/s to $\omega = 37$ rad/s. It is assumed that both the motor efficiency and the pump efficiency are one.

(1) Maximum torque T_m of the motor.
(2) The maximum swept volume $2\pi D_m$ of the variable displacement hydraulic motor.
(3) The inclined angle α of the swash plate corresponding to the maximum motor angular velocity $\omega = 37$ rad/s.

3.4 The formula expressing overall pump efficiency is given as follows:

$$\eta_p = \frac{1 - 0.75 \times 10^{-6} \frac{p}{\mu\omega}}{1.03 + 0.2 \times 10^4 \frac{\mu\omega}{p}}. \tag{3.140}$$

Calculate the maximum overall pump efficiency $\eta_{p\,\mathrm{max}}$ and the parameter $p/(\mu\omega)$ corresponding to the maximum efficiency.

3.5 Working fluid flows into a helical spline type rotary motor at the flow rate $q = 16$ L/min. Calculate the angular velocity of the rotary motor when the effective motor piston area A, the diameter d_p of helical spline pitch circle, the spline lead angle β and volumetric efficiency η_v of the motor are $A = 22.5$ cm^2, $d_p = 50$ mm, $\beta = 45°$ and $\eta_v = 0.75$, respectively.

3.6 A single vane type rotary motor produces the load with the moment inertia $J = 1.2$ kg $\cdot$ m^2 oscillatory due to the maximum differential pressure $\Delta p = 10$ MPa between the motor ports alternatively. Then the oscillation amplitude angle and angular frequency of the load are $\theta_p = \pi/3$ and $\omega = 6\pi$ rad/s, respectively. The overall efficiency and the torque efficiency of the motor are 0.8 and 0.9, respectively.

(1) Calculate the vane tip radius r_o when the vane root radius and the vane width are $r_i = 30\,\text{mm}$ and $b = 30\,\text{mm}$, respectively.

(2) Calculate the maximum power of the rotary motor in the oscillatory movement when the friction force is negligible.

3.7 As shown in Fig. 3.39, the piston plunger tip with equivalent mass m_e has penetrated at the velocity V_0 into the head cover chamber in the cushion mechanism. Then the plunger velocity changes to V_0/n within the time t_n when the plunger tip transfers at the distance x_n.

(1) Derive the formula expressing the mean velocity $V_{m1} = x_n/t_n$ in the cushion term when the cushion force f_1 is in proportion to the piston velocity.

(2) Derive the formula expressing the mean velocity $V_{m2} = x_n/t_n$ in the cushion term when the cushion force f_2 is in proportion to the square of the piston velocity.

(3) Compare graphically the mean velocity V_{m1} with V_{m2} depending on the variation of the rate n.

3.8 Working fluid flows into the valve chamber at the rate $Q = 240\,\text{L/min}$ through the ring-shaped spool port with the opening distance $x = 0.6\,\text{mm}$ and then flows out through the outlet port, as shown in Fig. 3.43(a). Calculate the differential pressure Δp at the inlet spool port and the steady axial force F_s acting on the spool. The angle of flow into the valve chamber is $\phi = 69°$, the spool diameter is $d = 16\,\text{mm}$, the flow coefficient is $C_d = 0.7$ at the inlet spool port and the working fluid density is $\rho = 860\,\text{kg/m}^3$.

3.9 Calculate the lift x of the poppet valve shown in Fig. 3.111 when the differential pressure at the valve port is $p_1 - p_2 = 0.8\,\text{MPa}$. The valve seat diameter is $d = 12\,\text{mm}$, the vertical angle of the poppet is $2\alpha = 90°$, the flow coefficient is $C_d = 0.7$, the spring constant is $k = 2 \times 10^4\,\text{N/m}$ and the spring deflection is $\delta_i = 2.25\,\text{mm}$ at the location when the valve is closed.

3.10 Calculate the ratio $\Delta p/p_s$ of the override pressure Δp to the setting pressure p_s in the directed relief valve shown in Fig. 3.51. The vertical angle of the poppet is $2\alpha = 60°$, the ratio of the lift x at the setting pressure p_s to the spring deflection x_0 at the location closing the

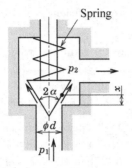

Fig. 3.111   Lift of poppet valve.

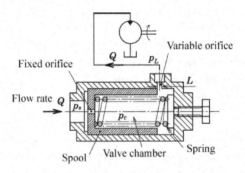

Fig. 3.112   Pressure compensated flow control valve.

poppet valve is $x/x_0 = 0.2$, the ratio of the poppet opening area A_x to the poppet port area S receiving the differential pressure is $A_x/S = 0.25$ and the flow coefficient in the port is $C_d = 0.7$.

3.11 Figure 3.112 shows a pressure compensated flow control valve to maintain a constant flow rate regardless of the load pressure fluctuations. The fluid flows into the valve chamber through the fixed orifice with the throttle area a_0 and then flows out of the valve chamber through the variable orifice. A is the spool area receiving the differential pressure, w is the width of the variable orifice throttle, L is the full opened length of the variable orifice, k is the spring constant, x is the spool displacement from the fully opened port position. The opening area of the variable orifice is denoted as $w(L - x)$, C_d and C_v are flow coefficients of the fixed and the variable orifice, respectively, and p_s, p_c and p_L are the fluid pressures at the locations shown in Fig. 3.112.

Answer the following questions relating to the case when the spring deflection δ at the fully open variable orifice position is denoted as $\delta = 10L$ and the angle of the fluid flow through variable orifice is $\phi = 90°$ for the spool axis.

(1) The force acting on the spool is balanced by the spool displacement $x = nL(n < 1)$. Derive the force balance equation of the spool using the flow rate Q passing through the valve. In addition, derive the formula expressing the flow rate Q passing through the fixed orifice and the formula governing the flow rate Q passing through the variable orifice.

(2) The spool force is balanced by the spool displacement $x = L/4$ on the condition that flow rate Q_{10} passes through the flow control valve. When the differential pressure $\Delta p_1 = p_s - p_L$ between the inlet and the outlet port of the valve changes to Δp_2 corresponding to the change of the load pressure p_L, the spool balancing displacement changes from the $x = L/4$ to $x = 3L/4$ automatically to maintain the constant flow rate through the valve. Then the flow rate Q_{10} changes slightly to Q_{20} in spite of the spool displacement. Calculate the ratio Q_{10}/Q_{20} and the ratio $\Delta p_1/\Delta p_2$ where the fixed orifice area is denoted as $a_0 = wL$ and both flow coefficients C_d and C_v are the same.

3.12 Flow rate Q_0 flows into the priority circuit through the port [4] in the priority valve shown in Fig. 3.64. Derive the spring deflection δ using the flow rate Q_0, fixed orifice area a, the flow coefficient C_d, the spring constant k, fluid density ρ, and the effective spool area A receiving the differential pressure.

3.13 Draw hydraulic logic circuits for the valves shown in Fig. 3.113 using the logic element symbols.

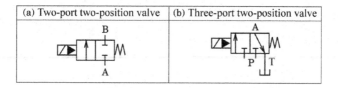

(a) Two-port two-position valve	(b) Three-port two-position valve

Fig. 3.113 Valve symbol.

3.14 The nitrogen gas pressure 12 MPa with the volume 14 L in the accumulator changes to 20 MPa due to a pressure change in the hydraulic circuit, and then changes from 20 MPa to 15 MPa when the working fluid in the accumulator is discharged into the hydraulic circuit within few seconds. Calculate the fluid volume discharged into the hydraulic circuit during the gas pressure change from 20 MPa to 15 MPa.

Chapter 4

Hydraulic Control Systems

The benefits that hydraulic control systems provide may be enhanced by the control of energy and the transformation of power to provide high-speed control response together with high power density (power to weight ratio). Moreover, high-precision hydraulic servo systems can be engineered by incorporating electronic computer control into the hydraulic systems. This chapter provides an introduction into basic control theory and its applications to electro hydraulic feedback control systems.

4.1 Configuration of the Components in Hydraulic Servo Systems

A servo system is defined as a feedback control system that controls mechanical positions, velocities or forces to continuously reduce the control error (i.e. deviation) between the output (i.e. control variable) and the reference input (i.e. command variable) varying in time. Basic components of electro hydraulic servo systems include:

- a command signal generator;
- a sensor to detect the control variable in the form of an electrical voltage signal;
- a servo amplifier generating electrical energy according to the signal voltage;
- a servo valve to transform electric energy into hydraulic energy;

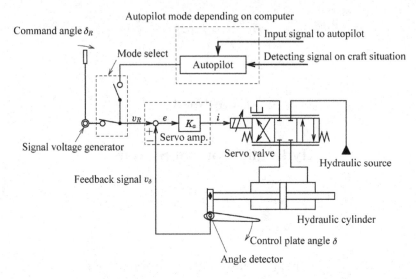

Fig. 4.1   Hydraulic aircraft steering servo system.

- a hydraulic actuator with load and
- the hydraulic circuit with the hydraulic power source.

A servo system that has high-precision, and that produces a large power and a high-speed response can be devised by incorporating a digital technology into the system.

An example of a servo system incorporating such features is given in Figure 4.1. The Figure illustrates a hydraulic aircraft steering servo system which can be switched into either auto pilot mode or manual steering mode. The steering input angle δ_R (command variable) and the control plate angle δ (control variable) are converted into voltage signals v_R and v_δ, respectively, by the differential transformers. The deviation voltage signal $e = v_R - v_\delta$ is input to the servo amplifier circuit, and is then transformed into the electric current $i = K_a e$ by the electric power amplifier circuit. This electric current is applied to the servo valve coil, and the servo valve supplies hydraulic fluid at a flow rate proportional to the magnitude of the electric current into the hydraulic cylinder that operates the control plate. This reduces the deviation e. Consequently, the angle δ of the control plate coincides with the input angle δ_R because the flow rate to the cylinder becomes zero at the time when the deviation voltage signal e reaches zero.

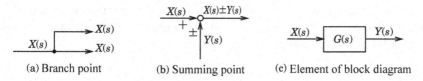

(a) Branch point (b) Summing point (c) Element of block diagram

Fig. 4.2 Rules of block diagram.

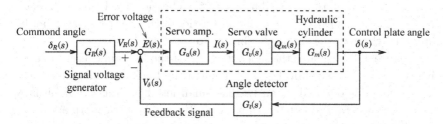

Fig. 4.3 Block diagram of a hydraulic aircraft steering servo system.

Such a process in a control system is typically represented by a block diagram so that the control characteristics may be analyzed. The principles of a block diagram are as follows:

(a) The block diagram illustrates the control process in the form of a signal flow chart. Arrows in a block diagram denote the direction of a passing control signal, and a letter marked with an arrow stands for a signal that is delivered. As shown in Fig. 4.2(a), the signal flow may be branched, and the branch point is indicated by a small black circle •. The magnitude of the signal does not change after the branch.
(b) As shown in Fig. 4.2(b), a small white circle ∘ indicates the summing point of the signals with the plus or the minus sign.
(c) A block in a block diagram denotes a control system element. As shown in Fig. 4.2(c), the signal $X(s)$ with the arrow reaching to the block stands for the input to the block whereas the signal $Y(s)$, the arrow starting from the block, stands for the output of the block. In the block diagram shown in Fig. 4.3, the next block does not affect the output of the previous one.

The transmission performances between the input and the output signals of the control elements are mathematically represented by a transfer function, which is usually defined in the block. The transfer function $G(s)$ is defined by the ratio of output $Y(s)$ to input $X(s)$ in the control element when $X(s)$ and $Y(s)$ respectively denote the Laplace transformation of the time variable input $x(t)$ and time variable output $y(t)$ in the state where all initial conditions are zero

$$G(s) = \frac{Y(s)}{X(s)}.$$ (4.1)

Let both $X(s)$ and $Y(s)$ be functions in the complex variable s in the Laplace transformation defined by Eq. (4.6) in Section 4.2. Utilizing Laplace transformation, the dynamical characteristics of the control system can be analyzed through the simple algebraic procedures.

Figure 4.3 illustrates the block diagram of the hydraulic aircraft steering servo system shown in Fig. 4.1. Though the Laplace transformation of a time variable denoted by an English small letter such as $x(t)$ is usually indicated by a capital letter such as $X(s)$, the Laplace transformation of a Greek letter such as $\delta(t)$ is given as $\delta(s)$ in this book.

In control engineering, a controlled variable denotes the physical quantity controlled as the system output, and a command variable denotes the desired quantity of the control variable supplied to the control system as the input. A feedback signal is the signal that is returned from the output end to the input end in order to compare the control variable with the command, as shown in Fig. 4.3. A control system with a feedback is referred to as a feedback control system or a closed loop control system.

In Fig. 4.3, the transfer functions $G_R(s), G_t(s)$ of the voltage generator to give the command angle and the sensor to detect the control plate angle are represented by

$$\frac{V_R(s)}{\delta_R(s)} = G_R(s), \quad \frac{V_\delta(s)}{\delta(s)} = G_t(s).$$ (4.2)

In addition, the transfer functions $G_a(s), G_v(s), G_m(s)$ of the servo amplifier, servo valve, and steering actuator are also expressed as follows:

$$\frac{I(s)}{E(s)} = G_a(s), \quad \frac{Q_m(s)}{I(s)} = G_v(s), \quad \frac{\delta(s)}{Q_m(s)} = G_m(s).$$ (4.3)

The entire transfer function enclosed with a chain line in Fig. 4.3, i.e. $\delta(s)/E(s)$ can be expressed as $G_a(s)G_v(s)G_m(s)$ because the product of the three equations in Eq. 4.3 gives the transfer function $\delta(s)/E(s) = G_a(s)G_v(s)G_m(s)$. In other words, the entire transfer function of control elements connected in series is represented by the product of transfer functions for each element. Therefore, the block diagram shown in Fig. 4.3 can

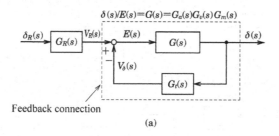

Feedback connection

(a)

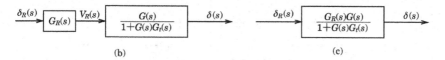

(b)   (c)

Fig. 4.4   Examples of block diagram reduction.

be reconfigured as Fig. 4.4(a). The block diagram enclosed with a chain line in Fig. 4.4(a) is called a feedback connection. In accordance with the rules of a block diagram, the signal relationships in the feedback connection are given as follows:

$$\left. \begin{array}{l} \delta(s) = E(s)G(s) \\ V_\delta(s) = \delta(s)G_t(s) \\ E(s) = V_R(s) - V_\delta(s) \end{array} \right\}. \tag{4.4}$$

Canceling algebraically variable signals $V_\delta(s)$ and $E(s)$ in Eq. (4.4) gives the relationship between the output $\delta(s)$ and input $V_R(s)$

$$\frac{\delta(s)}{V_R(s)} = \frac{G(s)}{1 + G(s)G_t(s)}. \tag{4.5}$$

The transfer function shown in Eq. (4.5) is called a closed loop transfer function, where $G(s)G_t(s)$ is referred to as the open loop transfer function in the block diagram enclosed with a chain line in Fig. 4.4(a). Recalling Eq. (4.5), the block diagram shown in Fig. 4.4(a) is rearranged as in Fig. 4.4(b), and consequently it can be replaced by Fig. 4.4(c).

In this manner, a complicated block diagram with feedback loops, etc. can be reduced by rearranging, in accordance with the rules of the block diagram algebra. Table 4.1 shows the equivalent transformation diagrams used in block diagram rearrangements.

Table 4.1   Rearrangements in partial block diagram.

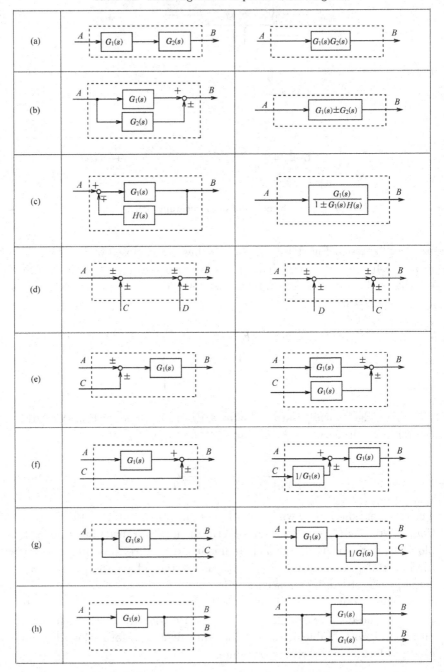

4.2 Basic Theory of Control Engineering

4.2.1 *Laplace Transformation and Inverse Laplace Transformation*

The equation relating output variable $y(t)$ to input variable $x(t)$ in an element is largely governed by a linear differential equation, depending on the property of the control element. A Laplace transformation can be used to solve a linear differential equation. When the single-valued time function $f(t)$ is continuous in every time interval in the range $t > 0$, the Laplace transformation $F(s)$ of a time function $f(t)$ is defined as follows:

$$F(s) = \mathcal{L}[f(t)] = \int_0^\infty f(t)e^{-st}dt, \qquad (4.6)$$

where s is a complex variable called a Laplace operator. The time function is transformed into an algebraic equation in a complex variable $s = \sigma + j\omega$, and $\mathcal{L}[f(t)]$ is the symbol indicating the Laplace transformation on a function $f(t)$. Considering the Laplace transformation of $f(t) = e^{-\alpha t}$ in the range $t > 0$, $\mathcal{L}[f(t)]$ is obtained as

$$F(s) = \int_0^\infty e^{-(s+\alpha)t}dt = \frac{1}{s+\alpha}. \qquad (4.7)$$

The definite integral $\int_0^\infty e^{-(s+\alpha)t}dt$ is obtained as $1/(s+\alpha)$ in the range of s where the real part of $-(s+\alpha)$ is minus, although it diverges when the real part takes positive values. In this manner, the function $f(t)$ can be transformed into the Laplace transformation $F(s)$ provided that there is the region of the complex variable s where the Laplace definite integration on $f(t)$ should converge.

The inverse Laplace transformation implies that a Laplace transformed function $F(s)$ is restored to the time function $f(t)$. The inverse Laplace transformation formula is given as

$$f(t) = \mathcal{L}^{-1}[F(s)] = \frac{1}{2\pi j} \int_{C-j\infty}^{C+j\infty} F(s)^{st}ds, \qquad (4.8)$$

where $j = \sqrt{-1}$ denotes an imaginary number, C is an arbitrary real constant and $\mathcal{L}^{-1}$ is the symbol denoting the inverse Laplace transformation.

Table 4.2   Laplace transformations of basic time variant functions.

$f(t)$	$F(s)$
(a) Unit impulse $\delta(t)$ $$+0 \leq t \leq T : \delta(t) = \lim_{T \to 0} \left(\frac{1}{T}\right)$$ $$t > T : \delta(t) = 0$$	1
(b) Unit step $u(t)$ $$t \geq +0 : u(t) = 1$$	$\dfrac{1}{s}$
(c) t^n	$\dfrac{n!}{s^{n+1}}$
(d) e^{-at}	$\dfrac{1}{s + \alpha}$
(e) $\dfrac{t^{n-1} e^{-\alpha t}}{(n-1)!}$	$\dfrac{1}{(s + \alpha)^n}$
(f) $\sin \omega t$	$\dfrac{\omega}{s^2 + \omega^2}$
(g) $\cos \omega t$	$\dfrac{s}{s^2 + \omega^2}$
(h) $A f_1(t) + B f_2(t)$	$A F_1(s) + B F_2(s)$
(i) $\dfrac{d^n f(t)}{dt^n}$	$s^n F(s) - s^{n-1} f(0) - s^{n-2} f^1(0) - \cdots$ $\cdots - s f^{n-2}(0) - f^{n-1}(0)$ where $f^k(0) = \left[d^k f(t)/dt^k\right]_{t=0}$
(j) $f(t - L), \quad (t < 0 : f(t) = 0)$	$e^{-Ls} F(s)$
(k) $e^{-\alpha t} f(t)$	$F(s + \alpha)$
(l) $\lim_{t \to 0} f(t)$	$\lim_{s \to \infty} s F(s)$
(m) $\lim_{t \to \infty} f(t)$	$\lim_{s \to 0} s F(s)$

Instead of recalling Eqs. (4.6) and (4.8), the Laplace transformation and the inverse Laplace transformation are usually obtained through algebraic operations in which the formulas defined in Tables 4.2 and 4.3 are applied. These list the Laplace transformations of the basic time functions and the theorems pertinent to the inverse Laplace transformation.

Let us consider a control element where the equation relating output variable $y(t)$ and input variable $x(t)$ is given as

$$\frac{d^3 y(t)}{dt^3} + 5 \frac{d^2 y(t)}{dt^2} + 9 \frac{dy(t)}{dt} + 5y(t) = 2x(t), \qquad (4.9)$$

Table 4.3   Partial fraction expansion theorem.

A transfer function is described as follows:

$$F(s) = \frac{N(s)}{D(s)}$$

where $N(s)$ and $D(s)$ respectively denote the m-degree polynomial and n-degree polynomial of Laplace operator s provided that $n \geq m$. Transfer function $F(s)$ can be expanded as shown in following conditions (a) and (b).

(a) For dominator $D(s)$ without multiple pole

$$F(s) = \frac{N(s)}{(s+s_1)(s+s_2)\cdots(s+s_i)\cdots(s+s_n)}$$

$$= \frac{K_1}{s+s_1} + \frac{K_2}{s+s_2} + \cdots + \frac{K_i}{s+s_i} + \cdots + \frac{K_n}{s+s_n}$$

where the constants $K_1, K_2, \ldots, K_i, \ldots, K_n$ are denoted as:

$$K_i = [F(s)(s+s_i)]_{s=-s_i}$$

Using Table 4.2(d) and (h), $f(t) = \mathcal{L}^{-1}[F(s)]$ is derived as:

$$f(t) = \mathcal{L}^{-1}\left[\sum_{i=1}^{n} \frac{K_i}{s+s_i}\right] = \sum_{i=1}^{n} K_i e^{-s_i t}$$

(b) For dominator $D(s)$ with multiple pole

$$F(s) = \frac{N(s)}{(s+\sigma)^r(s+s_1)(s+s_2)\cdots(s+s_i)\cdots(s+s_{n-r})}$$

$$= \frac{A_1}{(s+\sigma)^r} + \frac{A_2}{(s+\sigma)^{r-1}} + \cdots + \frac{A_k}{(s+\sigma)^{r-k-1}} + \cdots + \frac{A_r}{s+\sigma}$$

$$+ \frac{K_1}{s+s_1} + \frac{K_2}{s+s_2} + \cdots + \frac{K_i}{s+s_i} + \cdots + \frac{K_{n-r}}{s+s_{n-r}}$$

where A_1, A_2, A_k, K_i are denoted as

$$A_1 = [F(s)(s+\sigma)^r]_{s=-\sigma}, \quad A_2 = \left[\frac{d[F(s)(s+\sigma)^r]}{ds}\right]_{s=-\sigma}$$

$$A_k = \frac{1}{(k-1)!}\left[\frac{d^{k-1}[F(s)(s+\sigma)^r]}{ds^{k-1}}\right]_{s=-\sigma}, \quad K_i = [F(s)(s+s_i)]_{s=-s_i}$$

Using Table 4.2(d), (e) and (h), $f(t) = \mathcal{L}^{-1}[F(s)]$ is derived as

$$f(t) = \frac{A_1 t^{r-1} e^{-\sigma t}}{(r-1)!} + \frac{A_2 t^{r-2} e^{-\sigma t}}{(r-2)!} + \cdots + A_r e^{-\sigma t} + \sum_{i=1}^{n-r} K_i e^{-s_i t}$$

where the initial conditions are given by $d^2y(0)/dt^2 = dy(0)/dt = dy(0) = 0$ and $x(0) = 0$.

The Laplace transformation of Eq. (4.9) is obtained by virtue of items in Table 4.2(h) and (i)

$$s^3Y(s) + 5s^2Y(s) + 9sY(s) + 5Y(s) = 2X(s), \qquad (4.10)$$

where $X(s)$ and $Y(s)$ are Laplace transformations of $x(t)$ and $y(t)$, respectively. The transfer function $G(s)$ of the element is derived algebraically from Eq. (4.10) as follows:

$$G(s) = \frac{Y(s)}{X(s)} = \frac{2}{s^3 + 5s^2 + 9s + 5} = \frac{2}{(s+1)(s^2 + 4s + 5)}. \qquad (4.11)$$

Let us consider the indicial response that the step input $x(t) = u(t)$ is given as shown on the left side of Fig. 4.5. Since $X(s) = 1/s$ is given in Table 4.2(b), the output Laplace transformation $Y(s)$ becomes

$$\left. \begin{aligned} Y(s) = G(s)X(s) &= \frac{2}{s(s+1)(s+s_1)(s+s_2)} \\ -s_1 = -2 + j, \quad -s_2 &= -2 - j \end{aligned} \right\} . \qquad (4.12)$$

Applying the theorems shown in Table 4.3(a) to Eq. (4.12), Eq. (4.12) can be expressed in the form of a partial fraction. Consequently, the response

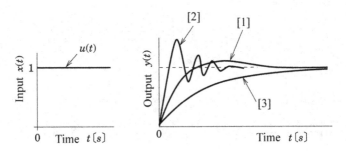

Fig. 4.5   Indicial responses.

of a control element $y(t) = \mathcal{L}^{-1}[G(s)/s]$ is given as

$$y(t) = \mathcal{L}^{-1}\left[\frac{K_1}{s} + \frac{K_2}{s+1} + \frac{K_3}{s+s_1} + \frac{K_4}{s+s_2}\right]$$

$$K_1 = \left[\frac{2s}{s(s+1)(s+s_1)(s+s_2)}\right]_{s=0} = \frac{2}{s_1 s_2} = \frac{2}{5}$$

$$K_2 = \left[\frac{2(s+1)}{s(s+1)(s+s_1)(s+s_2)}\right]_{s=-1}$$

$$= \frac{2}{-(-1+s_1)(-1+s_2)} = -1$$

$$K_3 = \left[\frac{2(s+s_1)}{s(s+1)(s+s_1)(s+s_2)}\right]_{s=-s_1}$$

$$= \frac{2}{-s_1(-s_1+1)(-s_1+s_2)} = \frac{3-j}{10}$$

$$K_4 = \left[\frac{2(s+s_2)}{s(s+1)(s+s_1)(s+s_2)}\right]_{s=-s_2}$$

$$= \frac{2}{-s_1(-s_2+1)(-s_2+s_1)} = \frac{3+j}{10}$$

$$(4.13)$$

Applying Table 4.2(b), (d) and (h) to Eq. (4.13), the inverse Laplace transformation of $Y(s)$ is derived as follows:

$$y(t) = \frac{2}{5} - e^{-t} + \frac{3-j}{10}e^{(-2+j)t} + \frac{3+j}{10}e^{(-2-j)t}. \qquad (4.14)$$

Substituting the mathematical formula $(e^{j\theta} + e^{-j\theta})/2 = \cos\theta$ and $(e^{j\theta} - e^{-j\theta})/(2j) = \sin\theta$ into Eq. (4.14) gives

$$y(t) = \frac{2}{5}\left[1 - \frac{5e^{-t}}{2} + \frac{\sqrt{10}}{2}e^{-2t}\sin(t + \tan^{-1}3)\right]. \qquad (4.15)$$

The output response behavior $y(t)$ can be derived accordingly using the Laplace transformation as long as the transfer functions of the system and the input are provided. The output response $y(t)$ to the step input $x(t) = u(t)$ is referred to as an indicial response, and the system that takes a constant steady value $y(\infty)$ for the indicial response is called a static system. For instance, the element defined by Eq. (4.11) is a static system because $y(\infty) = 2/5$.

The output response $y(t)$ changes from initial state to steady state through a transient behavior, which depends on the input that is applied and the property of the element. The transient behaviors are referred to as dynamic characteristics, and are often evaluated by the speed of response and the stability of indicial responses. Figure 4.5 illustrates the transient states in the indicial responses. With regard to the response speed only, the response wave [2] may be superior to others because it reaches the target value faster than the waves [1] and [3]. As regards the stability of response, however, the wave [2] appears to be inferior because of its transient oscillatory behavior. In respect of the response speed and stability, the response wave [1] appears to have the most favorable features. Table 4.4 summarizes the evaluation indexes to be used in dynamic characteristics.

Table 4.4 Transient characteristics for indicial response.

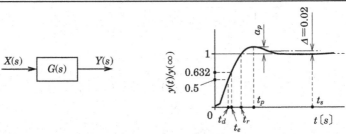

Configuration of the step response for a static system gives the evaluation index of the response speed and stability. Where $y(\infty)$ denotes value of y at steady state condition.

(a) Evaluation index for speed response

    (1) Delay time t_d: Response time reach at $0.5y(\infty)$
    (2) Rise time t_r: Response time to reach at $y(\infty)$ or $0.98y(\infty)$
    (3) Peak time t_p: Response time to reach at the first peak of the overshoot
    (4) Equivalent time constant t_e: Response time to reach at $0.632y(\infty)$

(b) Evaluation index for stability

    (5) Maximum overshoot a_p: $a_p = \dfrac{y(t_p) - y(\infty)}{y(\infty) - y(0)}$

(c) Evaluation index for response speed considering stability

    (6) Settling time t_s: First response time staying a range between $0.98y(\infty)$ and $1.02y(\infty)$

4.2.2 *Dynamic Characteristics of Linear Systems*

4.2.2.1 *Transient Response of Linear Systems*

Let us consider a system whose output responses to the inputs $x_1(t)$ and $x_2(t)$ are denoted by $y_1(t)$ and $y_2(t)$, respectively. The system is called linear as long as it produces an output response $y_1(t) + y_2(t)$ to the input $x_1(t) + x_2(t)$. In other words, the system is linear if, in the relationship between the input and output, the superposition principle applies to its response. Although analyses based on Laplace transformations are restricted to linear systems, the majority of real physical systems may be considered to be governed by linear mathematical models as long as the variability of the relevant parameters is limited. Moreover, an approximate equation linearizing the nonlinear system may be used when investigating a control object or an element, since the control system is designed to handle minor fluctuations around a controlled equilibrium state. The physical relationships between the input $x(t)$ and output $y(t)$ in a linear system may be expressed by the following differential equation

$$\frac{d^n y}{dt^n} + a_1 \frac{d^{n-1} y}{dt^{n-1}} + a_2 \frac{d^{n-2} y}{dt^{n-2}} + \cdots + a_{n-1} \frac{dy}{dt} + a_n y$$

$$= b_0 \frac{d^m x}{dt^m} + b_1 \frac{d^{m-1} x}{dt^{m-1}} + \cdots + b_{m-1} \frac{dx}{dt} + b_m x, \qquad (4.16)$$

where n and m are integers and $n \geq m$. Though the solution of Eq. (4.16) can be derived for the various initial conditions, all the initial values $x(0), dx(0)/dt, dx^2(0)/dt^2 \cdots, y(0), dy(0)/dt, \ldots$ may be regarded as zero because the input and output variables $x(t)$ and $y(t)$ are treated as variables fluctuating from an equilibrium point. Applying the expressions in Table 4.2(h) and (i) to Eq. (4.16) gives the transfer function of the system as follows:

$$G(s) = \frac{Y(s)}{X(s)} = \frac{b_0 s^m + b_1 s^{m-1} + \cdots + b_{m-1} s + b_m}{s^n + a_1 s^{n-1} + a_2 s^{n-2} + \cdots + a_{n-1} s + a_n}. \qquad (4.17)$$

The equation that brings the denominator in the transfer function to zero is called a characteristic equation. The characteristic equation for Eq. (4.17) is given as

$$s^n + a_1 s^{n-1} + a_2 s^{n-2} + \cdots + a_{n-1} s + a_n = 0. \qquad (4.18)$$

When characteristic roots of Eq. (4.18) include the real roots $-s_i = -\alpha_i$ (where $i = 1, 2, \ldots, p$) and the pairs of the conjugate complex roots

$-s_k = -\sigma_k \pm j\omega_k$ (where $k = 1, 2, \ldots, q$), the transfer function $G(s)$ shown in Eq. (4.18) can be rearranged as follows:

$$G(s) = \frac{Y(s)}{X(s)}$$

$$= \frac{b_0 s^m + b_1 s^{m-1} + \cdots + b_{m-1} s + b_m}{(s+\alpha_1)\cdots(s+\alpha_p)(s+\sigma_1+j\omega_1)(s+\sigma_1-j\omega_1)\cdots(s+\sigma_q+j\omega_q)(s+\sigma_q-j\omega_q)}.$$

$$(4.19)$$

Applying Table 4.3(a) to Eq. (4.19), the Laplace transformation $Y(s) = G(s)/s$ of the output to the step input $X(t) = 1/s$ is given as

$$Y(s) = \frac{G(s)}{s} = \frac{K_0}{s} + \sum_{i=1}^{p} \frac{K_i}{s+\alpha_i} + \sum_{k=1}^{q} \left(\frac{C_k}{s+\sigma_k+j\omega_k} + \frac{\bar{C}_k}{s+\sigma_k-j\omega_k} \right),$$

$$(4.20)$$

where K_0, K_i, C_k and $\bar{C}_k$ are the constants established by the theorem in Table 4.3(a). Real characteristic roots $-s_i = -\alpha_i$, (where $i = 1, 2, \ldots, p$) and conjugate complex characteristic roots $-s_k = -\sigma_k \pm j\omega_k$ (where $k = 1, 2, \ldots, q$) are called the poles of the transfer function $G(s)$, whereas complex variables s bringing the numerator of $G(s)$ to zero are called the zeros of $G(s)$. Applying expressions Table 4.2(b) and (d) to Eq. (4.20), the indicial response $y(t)$ can be derived as follows:

$$y(t) = \mathcal{L}^{-1}\left[\frac{G(s)}{s}\right] = K_0 + \sum_{i=1}^{p} K_i e^{-\alpha_i t} + \sum_{k=1}^{q} e^{-\sigma_k t}(C_k e^{j\omega_k t} + \bar{C}_k e^{-j\omega_k t}).$$

$$(4.21)$$

Substituting the expansion $e^{\pm j\omega_k t} = \cos\omega_k t \pm j\sin\omega_k t$ to Eq. (4.21) gives

$$y(t) = K_0 + \sum_{i=1}^{p} K_i e^{-\alpha_i t} + \sum_{k=1}^{q} A_k e^{-\sigma_k t} \sin(\omega_k t + \phi_k), \qquad (4.22)$$

where A_k and ϕ_k are real constants. The second term $\sum_{i=1}^{p} K_i e^{-\alpha_i t}$ on the right side of Eq. (4.22) indicates the exponential attenuating behavior when $-\alpha_i < 0$ and the exponential divergent behavior when $-\alpha_i > 0$. The third term $\sum_{k=1}^{q} A_k e^{-\sigma_k t} \sin(\omega_k t + \phi_k)$ represents the damped oscillatory behavior for $-\sigma_k < 0$ and the divergent oscillatory behavior for $-\sigma_k > 0$. Consequently, the system remains stable provided all real parts of the characteristic roots have a negative value. In the case of stable systems, it appears that the entire dynamic behavior is dominated by the term with the smallest value in $|-\alpha_i|$ or $|-\sigma_k|$ because the terms that have larger values in $|-\alpha_i|$ and $|-\sigma_k|$ will be attenuated more quickly and then disappear. It is worth studying the response performance of the first and the second order lag system, because a linear system's response is not only represented by the wave superimposing some of the responses of the first and the second order lag system, but it can often be approximated by just a first or second order lag system.

4.2.2.2 *Response Characteristics of the First Order Lag System*

The transfer function of a first order lag system is given as

$$G_1(s) = \frac{Y(s)}{X(s)} = \frac{K}{Ts + 1}, \tag{4.23}$$

where $X(s)$ and $Y(s)$ denote the Laplace transformation of the input $x(t)$ and output $y(t)$ in the system, respectively. Recalling the section of Table 4.3(a) with relevance to Eq. (4.23), the indicial response of the first order lag system may be derived as follows:

$$y(t) = \mathcal{L}^{-1}\left[\frac{K}{s} - \frac{K}{s + (1/T)}\right] = K[1 - e^{-(1/T)t}]. \tag{4.24}$$

As shown in Eq. (4.24), the magnitude of the indicial response $y(t)$ is proportional to K, where K is called the gain constant. Figure 4.6 shows indicial response $y(t)$ of the first order lag system for the dimensionless time t/T. Response $y(T)$ at time $t = T$ takes the value corresponding to $0.632y(\infty) = 0.632K$. The factor T is referred to as a time constant, and is an index indicating the response speed for the first order lag system.

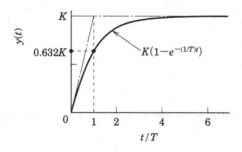

Fig. 4.6 Indicial response of a first order lag system.

4.2.2.3 *Response of a Standard Second Order Lag System*

A second order lag system has the transfer function in which the fractions in the third term of Eq. (4.20) are reduced to a common denominator

$$\left(\frac{C_k}{s + \sigma_k + j\omega_k} + \frac{\bar{C}_k}{s + \sigma_k - j\omega_k} \right) = \frac{b_1 s + b_2}{s^2 + 2\sigma_k s + \sigma_k^2 + \omega_k^2}, \qquad (4.25)$$

where b_1 and b_2 are real constants. The damping performance of a second order lag system depends on the real parts $-\sigma_k$ of the characteristic roots, and the response speed mainly depends on the imaginary parts ω_k of the characteristic roots, as shown in Eq. (4.22). However, the first term $b_1 s$ in the numerator of Eq. (4.25) hardly affects the dynamic response except for the very short period in the neighborhood of the time $t = 0$. Therefore, the dynamic characteristics of a second order lag system are usually assumed to be those of a standard second order lag system, and the term $b_1 s$ is neglected, for simplicity. The transfer function of the standard second order lag system is defined as

$$G_2(s) = \frac{K\omega_n^2}{s^2 + 2\zeta\omega_n s + \omega_n^2}. \qquad (4.26)$$

The factor ζ is called a damping coefficient or a damping ratio, and ω_n is referred to as an undamped natural angular frequency. Characteristic roots s_1 and s_2 for transfer function $G_2(s)$ are denoted as follows:

$$\left. \begin{array}{l} \zeta > 1 : -s_1 = -\zeta\omega_n + \omega_n\sqrt{\zeta^2 - 1}, \quad -s_2 = -\zeta\omega_n - \omega_n\sqrt{\zeta^2 - 1} \\[2mm] \zeta = 1 : -s_1 = -s_2 = -\omega_n \\[2mm] 0 < \zeta < 1 : -s_1 = -\zeta\omega_n + j\omega_n\sqrt{1 - \zeta^2}, \\[2mm] \qquad\qquad -s_2 = -j\omega_n - j\omega_n\sqrt{1 - \zeta^2} \end{array} \right\}.$$

$$(4.27)$$

Applying the same procedure as that used to solve Eq. (4.12), the indicial response $y(t) = \mathcal{L}^{-1}\left[\frac{K\omega_n^2}{s(s+s_1)(s+s_2)} \right]$ is derived as follows:

For $\zeta > 1$:

$$\left. \begin{array}{c} y(t) = K\left[1 - \dfrac{e^{-\zeta\omega_n t}}{\sqrt{\zeta^2 - 1}} \sinh(\omega_n\sqrt{\zeta^2 - 1}\,t + \phi) \right] \\[4mm] \phi = \tanh^{-1}\dfrac{\sqrt{\zeta^2 - 1}}{\zeta} \end{array} \right\}. \qquad (4.28)$$

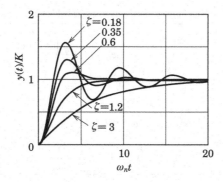

Fig. 4.7   Indicial responses of a standard second order lag system.

For $\zeta = 1$:

$$y(t) = K[1 - e^{-\omega_n t}(\omega_n t + 1)].\qquad(4.29)$$

For $\zeta < 1$:

$$\left.\begin{array}{c} y(t) = K\left[1 - \dfrac{e^{-\zeta\omega_n t}}{\sqrt{1 - \zeta^2}}\sin(\omega_n\sqrt{1 - \zeta^2}\,t + \phi)\right] \\[4mm] \phi = \tan^{-1}\dfrac{\sqrt{1 - \zeta^2}}{\zeta} \end{array}\right\}.\qquad(4.30)$$

Given the specified damping coefficients, indicial responses $y(t)/K$ for the dimensionless time $\tau = \omega_n t$ can be obtained from Eqs. (4.28), (4.29) and (4.30), and they are shown in Fig. 4.7.

The magnitude of response $y(t)$ is proportional to the gain constant K. The stability depends on the damping coefficient, whereas the response speed depends on the parameter $\omega_n\sqrt{1 - \zeta^2}$ which is called the natural angular frequency. The damping performance of these responses becomes as follows:

- For $\zeta < 0$: the system is unstable,
- For $\zeta = 0$: the system becomes just oscillatory,
- For $0 \leq \zeta < 1$: the system response is under-damped,
- For $\zeta = 1$: the system response is critically damped.

Considering the both stability and response speed, the damping coefficient $\zeta = 0.7$ is usually regarded as the optimum. For $\zeta > 1$, the system response is over damped. Table 4.5 shows the response characteristics of a standard second order lag system calculated from Eq. (4.30) in case of $\zeta < 1$.

Table 4.5   Response characteristics of a standard second order lag system.

(a) Rise time : $y(t_r) = 1$	$t_r = \dfrac{1}{\omega_n \sqrt{1-\zeta^2}} \left(\pi - \tan^{-1} \dfrac{\sqrt{1-\zeta^2}}{\zeta} \right)$
(b) Peak time	$t_p = \dfrac{\pi}{\omega_n \sqrt{1-\zeta^2}}$
(c) Maximum overshoot	$a_\rho = \exp \left(-\dfrac{\pi\zeta}{\sqrt{1-\zeta^2}} \right)$
(d) Settling time	$\Delta = 2\% : t_s = \dfrac{3.9}{\zeta\omega_n}, \quad \Delta = 5\% : t_s = \dfrac{3}{\zeta\omega_n}$

4.2.2.4 *Approximated Transfer Function of a System*

Since indicial response $y(t)$ for $\zeta \geq 1$ asymptotically tends to the steady value without an overshoot, it may be approximated by the indicial response $y_e(t)$ of the first order lag system with an equivalent time constant T_e corresponding to the value of $y(T_e) = 0.632K$

$$\frac{Y_e(s)}{X(s)} = \frac{K}{T_e s + 1}. \tag{4.31}$$

The dimensionless equivalent time constant $\tau_e = \omega_n T_e$ satisfying the requirement $y(\tau_e) = 0.632K$ for the specified values of ζ can be numerically calculated because Eq. (4.28) is regarded as a function of variables $\tau = \omega_n t$ and ζ. The relationship between ζ and $\omega_n T_e$ for $\zeta \geq 1$ is obtained as follows:

$$T_e \approx \frac{2\zeta}{\omega_n}. \tag{4.32}$$

Though the value of $\omega_n T_e$ derived from Eq. (4.29) for $\zeta = 1$ is about 7% larger than the value derived from Eq. (4.32), it approaches the value obtained from Eq. (4.32) as the damping coefficient ζ increases.

Using the dimensionless time τ and Eq. (4.32), the indicial response $y(t)$ in Eq. (4.28) is written as

$$y(\tau) = K \left[1 - \frac{1}{\sqrt{\zeta^2 - 1}} e^{-2\zeta^2 \tau} \sinh(2\zeta \sqrt{\zeta^2 - 1}\tau + \phi) \right], \tag{4.33}$$

$$\tau = \frac{t}{T_e}. \tag{4.34}$$

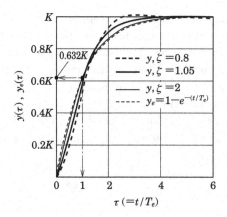

Fig. 4.8 Comparison between the curve $y_e(\tau)$ and curves $y(\tau)$.

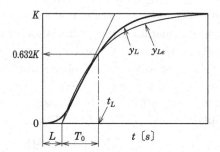

Fig. 4.9 Indicial response curve of the system.

The indicial response approximated by $y_e(\tau) = \mathcal{L}^{-1}[Y_e(s)/s]$ is denoted as

$$y_e(\tau) = K(1 - e^{-\tau}). \tag{4.35}$$

Figure 4.8 reveals that the differences between the indicial response $y_e(\tau)$ and $y(\tau)$ are small. Therefore, a standard second order lag system for $\zeta \geq 1$ is often approximated by a first order lag system with an equivalent time constant T_e.

The indicial response of a system occasionally exhibits an S-shaped curve with the inactive rising zone around $t = 0$ such as the curve $y_L(t)$ in Fig. 4.9. Thus, it is often approximated by the curve $y_{Le}(t)$ defined by Eq. (4.36).

$$\left. \begin{array}{l} 0 \leq t \leq L : y_{Le}(t) = 0 \\ t > L : y_{Le}(t) = K[1 - e^{-(1/T_0)(t-L)}] \end{array} \right\}. \tag{4.36}$$

Equation (4.36) expresses an indicial response of a first order lag system occurring after the pure dead time L has passed. In the approximated transfer function, the dead time L is estimated by the time when the straight line tangent to the curve $y_L(t)$ at the inflection point intersects the time axis, and then time constant T_0 is obtained by subtracting the pure dead time L from the response time t_L at $y_L(t_L) = 0.632K$.

Recalling Table 4.2(j) and by virtue of Eq. (4.36), the transfer function $G_L(s)$ corresponding to the indicial response $y_{L_e}(t)$ is given as

$$G_L(s) = \frac{Y_{Le}(s)}{X(s)} = \frac{Ke^{-Ls}}{T_0s + 1}. \tag{4.37}$$

4.2.3 Steady-state Errors

Figure 4.10 shows a block diagram of a feedback control system which regulates the control variable $c(t)$ so that it should agree with the input $r(t)$ by using the control operation with a deviation of $e(t) = r(t) - c(t)$. However, if the steady state control variable $c(\infty)$ does not agree with the steady state input $r(\infty)$ due to the nature of the input $r(t)$ or of the open loop transfer function $G(s)H(s)$, then the error $e(\infty) = r(\infty) - c(\infty)$ is referred to as a steady-state error. The block diagram rules being applied to Fig. 4.10, the Laplace transformation $E(s)$ of error signal $e(t)$ is given as

$$E(s) = R(s) - C(s) = R(s)\frac{1 + G(s)[H(s) - 1]}{1 + G(s)H(s)}. \tag{4.38}$$

Applying Table 4.2(m) to Eq. (4.38) gives

$$e(\infty) = \lim_{s \to 0} sE(s) = \lim_{s \to 0} \frac{sR(s)\{1 + G(s)[H(s) - 1]\}}{1 + G(s)H(s)}. \tag{4.39}$$

The transfer function $H(s)$ of a feedback element yields the following condition:

$$h(\infty) = \lim_{s \to 0} H(s) = 1. \tag{4.40}$$

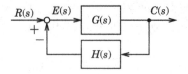

Fig. 4.10   Control system.

Thus, the steady-state error $e(\infty)$ becomes

$$e(\infty) = \lim_{s \to 0} sE(s) = \lim_{s \to 0} \frac{sR(s)}{1 + G(s)}. \qquad (4.41)$$

The steady-state error $e(\infty) = e_p$ for the step input $r(t) = r_0 u(t)$ is called a steady-state position error. Substituting $R(s) = \mathcal{L}[r_0 u(t)] = r_0/s$ into Eq. (4.41) gives

$$e_p = \lim_{s \to 0} \frac{r_0}{1 + G(s)} = \frac{r_0}{1 + \lim_{s \to 0} G(s)}. \qquad (4.42)$$

The steady-state position error e_p depends on the constant K_p, where $[G(s)]_{s=0} = K_p$. The constant K_p is referred to as a steady-state position error constant.

The steady-state error $e(\infty) = e_v$ for a constant speed input $dr(t)/dt = r_v u(t)$ at $\dot{r}(0) = [dr/dt]_{t=0} = 0$ is called a steady-state velocity error. Since the Laplace transformation $\mathcal{L}[r(t)]$ is obtained as $R(s) = r_v/s^2$ by recalling Table 4.2(i) for $dr(t)/dt = r_v u(t)$, substituting $R(s) = r_v/s^2$ into Eq. (4.41) gives

$$e_v = \lim_{s \to 0} \frac{r_v}{s[1 + G(s)]} = \frac{r_v}{\lim_{s \to 0} s\,G(s)}. \qquad (4.43)$$

Therefore, the steady-state velocity error e_v depends on the constant K_v, where $[sG(s)]_{s=0} = K_v$, and it is called a steady-state velocity error constant.

The universal open loop transfer function $G(s)H(s)$ in Fig. 4.10 may be defined as

$$G(s)H(s) = \frac{K(1 + \beta_1 s + \beta_2 s^2 + \cdots + \beta_m s^m)}{s^q(1 + \alpha_1 s + \alpha_2 s^2 + \cdots + \alpha_p s^p)}, \qquad (4.44)$$

where $q + p \geq m$, $q \geq 0$, and m, p and q are arbitrary positive integers including zero. Equation (4.44) gives

$$\lim_{s \to 0} G(s)H(s) = \lim_{s \to 0} G(s) = \lim_{s \to 0} \frac{K}{s^q}. \qquad (4.45)$$

Equation (4.45) implies that the exponent q in Eq. (4.45) has a significant affect on the steady-state error. Therefore, control systems may be classified depending on the number of q, and a control system may be called a type-zero system, a type-one system or a type-two system in accordance with the number of $q = 0, 1$ and 2. The steady state position error e_p and the steady-state velocity error e_v for $q = 0, 1$ and 2 are summarized in Table 4.6.

Table 4.6   Steady-state position errors and steady-state velocity errors.

Type of control system	$r = r_0 u(t)$		$\dfrac{dr}{dt} = r_v u(t)$	
	K_p	e_p	K_v	e_v
Type-zero ($q = 0$)	K	$r_0/(1 + K)$	0	∞
Type-one ($q = 1$)	∞	0	K	r_v/K
Type-two ($q = 2$)	∞	0	∞	0

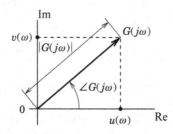

Fig. 4.11   Vector $G(j\omega)$.

4.2.4 Frequency Responses

4.2.4.1 Frequency Transfer Function and Frequency Response

A frequency transfer function $G(j\omega)$ is obtained by substituting $s = j\omega$ into a transfer function $G(s)$. The frequency transfer function $G(j\omega)$ involves a real function $u(\omega)$ and an imaginary function $jv(\omega)$ in angular frequency ω rad/s. Figure 4.11 shows a complex plane with a real axis Re and an imaginary axis Im. The frequency transfer function may be represented on the complex plane by the vector of a gain $|G(j\omega)|$ which denotes the vector magnitude, and the phase angle $\angle G(j\omega)$ which denotes the inclined angle of the vector to the real axis. The vector $G(j\omega)$ changes with variations of angular frequency ω, and it is expressed as

$$G(j\omega) = u(\omega) + jv(\omega)$$
$$= |G(j\omega)|[\cos \angle G(j\omega) + j \sin \angle G(j\omega)] = |G(j\omega)|e^{j\angle G(j\omega)}, \quad (4.46)$$

where the gain $|G(j\omega)|$ and phase angle $\angle G(j\omega)$ are represented by

$$|G(j\omega)| = \sqrt{u^2(\omega) + v^2(\omega)}, \quad (4.47)$$

$$\angle G(j\omega) = \tan^{-1} \frac{v(\omega)}{u(\omega)}. \quad (4.48)$$

Fig. 4.12 Frequency response.

An element used in a control system is generally treated as a stable system. Given the sinusoidal input $x(t)$ to a stable system as shown in Fig. 4.12, the steady-state response output $y(t)$ is given as

$$x(t) = A_i \sin \omega t, \qquad (4.49)$$

$$y(t) = A_o \sin(\omega t + \phi). \qquad (4.50)$$

The phase angle ϕ and amplitude A_o of the frequency response can be obtained directly using the frequency transfer function $G(j\omega)$ of the stable system

$$\frac{A_o}{A_i} = |G(j\omega)| = \sqrt{u^2(\omega) + v^2(\omega)}, \qquad (4.51)$$

$$\phi = \angle G(j\omega) = \tan^{-1} \frac{v(\omega)}{u(\omega)}. \qquad (4.52)$$

Recalling Table 4.2(f), the Laplace transformation of the output $y(t)$ is written as $Y(s) = X(s)G(s) = G(s)\omega/(s^2 + \omega^2)$. Equations (4.50), (4.51) and (4.52) can be derived by operating the inverse Laplace transformation of $Y(s)$ since all real parts of poles of the transfer function given in Eq. (4.19) are minus-valued in a stable system.

The steady-state response given by Eq. (4.50) is called a frequency response. The behavior of the gain $|G(j\omega)|$ and phase angle $\angle G(j\omega)$ in response to frequency variations are called the frequency characteristics. These exhibit not only the dynamic characteristics of the system with $G(s)$, but also provide a technical method to synthesize the feedback control system with an open loop transfer function $G(s)$.

4.2.4.2 *Bode Diagram*

A Bode diagram represents the frequency characteristics of the system by means of two graphs in which the horizontal axis indicates the angular frequency. One is the logarithmic curve of gain against the angular frequency variation, and the other is the curve of phase angle against the angular frequency variation. The logarithmic gain of $G(\omega)$ in decibels is defined as follows:

$$20 \log \left(\frac{A_o}{A_i} \right) = 20 \log |G(j\omega)| \ [\text{dB}]. \tag{4.53}$$

The Bode diagrams are plotted on a semi-log graph with a log scale for the horizontal axis denoting the angular frequency, and linear scales for the vertical axis denoting both the logarithmic gain and the phase angle.

Let us now consider the frequency transfer function $G(j\omega) = G_1(j\omega)G_2(j\omega) \cdots G_n(j\omega)$ involving the product of several frequency transfer functions. The frequency transfer function $G(\omega)$ being defined by the vector description in Eq. (4.46), the frequency transfer function $G(j\omega)$ can be written as

$$G(j\omega) = |G_1(j\omega)||G_2(j\omega)| \cdots |G_n(j\omega)|e^{j[\angle G_1(j\omega)+\angle G_2(j\omega)+\cdots+\angle G_n(j\omega)]}. \tag{4.54}$$

Therefore, the logarithmic gain of $G(j\omega)$ and the phase angle $\angle G(j\omega)$ are expressed as

$$20 \log |G(j\omega)|$$
$$= 20 \log |G_1(j\omega)| + 20 \log |G_2(j\omega)| + \cdots + 20 \log |G_n(j\omega)| \ [\text{dB}]. \tag{4.55}$$

$$\angle G(j\omega) = \angle G_1(j\omega) + \angle G_2(j\omega) + \cdots + G_n(j\omega). \tag{4.56}$$

Equations (4.55) and (4.56) imply that the Bode diagram of $G(j\omega)$ may be constructed by graphically adding the individual Bode diagrams $G_1(j\omega), G_2(j\omega), \ldots, G_n(j\omega)$ for the vertical direction. For instance, the Bode diagram of $G(j\omega) = K/j\omega$ may be depicted by adding graphically both Bode diagrams of $G_1(j\omega) = 1/(j\omega)$ and $G_2(j\omega) = K$ in the vertical direction. The Bode diagram of $G_1(j\omega) = 1/(j\omega)$ consists of a gain straight line of $-20 \log \omega$ [dB] with the slope -20 dB/decade which passes through the point $20 \log |G_1(j\omega)| = 0$ dB at $\omega = 1$ rad/s together with a straight phase line for the phase angle $\angle G_1(j\omega) = \tan^{-1}(-\omega/0) = -90°$ for any angular frequency ω. As regards the function $G_2(\omega) = K$, the logarithmic gain is denoted by $20 \log K$ [dB], and the phase angle $\angle G_2(j\omega) = \tan^{-1}(0/\omega)$ is zero for any frequency.

Therefore, the dashed straight line in Fig. 4.13 denotes the gain Bode diagram of $G(j\omega) = K/(j\omega)$ which corresponds to the line obtained by parallel-shifting the solid line of $20 \log |G_1(j\omega)| = -20 \log \omega$ as far as $20 \log K$ in the vertical direction. The phase angle in the Bode diagram of $\angle G(j\omega) = \angle K/(j\omega)$ equals to the phase angle Bode diagram of $G_1(j\omega) = 1/(j\omega)$ because of $\angle G_2(\omega) = 0$.

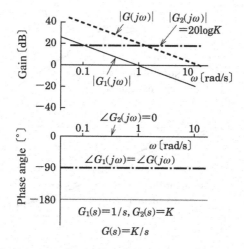

Fig. 4.13   Bode diagram of $G(j\omega) = K/(j\omega)$.

(a) *Bode diagram of the first order lag element*

Recalling Eqs. (4.23), (4.46), (4.47) and (4.48), the frequency transfer function of a first order lag system with unit gain constant $K = 1$ is given as follows:

$$G_1(j\omega) = \frac{1}{1 + jT\omega} = \frac{e^{j\phi}}{\sqrt{1 + (\omega T)^2}}, \quad \phi = \tan^{-1}(-\omega T). \qquad (4.57)$$

The gain $|G_1(j\omega)|$ and phase angle $\phi = \angle G_1(j\omega)$ are expressed as

$$\left.\begin{array}{l} 20\log|G_1(j\omega)| = -10\log[1 + (\omega T)^2] \text{ [dB]} \\ \angle G_1(j\omega) = \tan^{-1}(-\omega T) \end{array}\right\} . \qquad (4.58)$$

Figure 4.14 illustrates the Bode diagram of the first order lag system $G_1(j\omega)$ plotted against the dimensionless angular frequency ωT. Equation (4.58) implies that the logarithmic gain of $G_1(j\omega)$ nearly equals zero dB in the frequency range $\omega T \ll 1$. However, it is $-20\log 2 \approx -3$ dB at frequency $\omega T = 1$, and the gain curve can be approximated by the line of $-20\log(\omega T)$ for $\omega T \gg 1$. Therefore, the gain Bode diagram of $G_1(j\omega)$ can be approximated by two asymptotes. One of them is a straight line on zero dB for the angular frequency range $0 < \omega < 1/T$, while that for $\omega > 1/T$ is a straight line with a slope -20 dB/decade starting from zero dB at the frequency

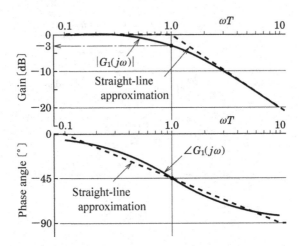

Fig. 4.14   Bode diagram of the first order lag system.

$\omega_b = 1/T$. Thus, the maximum error in the approximated Bode diagram of gain is 3 dB at $\omega = 1/T$. Consequently the gain and phase angle at $\omega = \omega_b = 1/T$ are as follows:

$$\left.\begin{aligned} 20\log|G_1(j\omega_b)| &= -3 \text{ dB} \\ \angle G_1(j\omega_b) &= -45° \end{aligned}\right\}. \tag{4.59}$$

The approximated gain Bode diagram is called a straight-line approximation and the frequency $\omega_b = 1/T$ is referred to as the break point angular frequency. The phase angle Bode diagram of the first order lag system may be also expressed approximately by three straight-lines. Namely, the first straight-line is kept at the zero degree in angular frequency range $\omega < 0.1/T$. The second straight line is inclined at $-45°$/decade passing through $\angle G_1(j\omega_b) = -45°$ at the break frequency $\omega_b = 1/T$ in the angular frequency range $0.1/T < \omega < 10/T$. The third straight-line is kept at $-90°$ for $\omega > 10/T$. The maximum error in the approximated phase angle diagram is about $7°$.

Figure 4.15 illustrates the Bode diagram of $G(j\omega) = 1/[j\omega(0.5j\omega + 1)]$ plotted by solid lines, the Bode diagram of $G_1(j\omega) = 1/(j\omega)$ graphed with dotted lines and the Bode diagram of $G_2(j\omega) = 1/(0.5j\omega + 1)$ plotted with chain lines. It is well established that the Bode diagram of $G(j\omega)$ can be plotted by adding graphically the Bode diagram of $G_1(j\omega)$ to the Bode diagram of $G_2(j\omega)$.

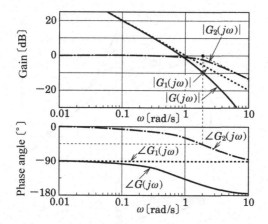

Fig. 4.15   Bode diagram of $G(j\omega) = 1/[j\omega(0.5j\omega + 1)]$.

(b) *Bode diagram of a standard second order lag system*

Recalling Eq. (4.26), the frequency transfer function $G_2(j\omega)$ of the standard second order lag system with $K = 1$ is defined as

$$G_2(j\omega) = \frac{1 - (\omega/\omega_n)^2 - j2\zeta(\omega/\omega_n)}{[1 - (\omega/\omega_n)^2]^2 + 4\zeta^2(\omega/\omega_n)^2}, \tag{4.60}$$

$$\left.\begin{array}{l} 20\log|G_2(j\omega)| = -10\log\{[1 - (\omega/\omega_n)^2]^2 + (2\zeta\omega/\omega_n)^2\} \ [\text{dB}] \\[2mm] \angle G_2(j\omega) = -\tan^{-1}\dfrac{2\zeta\omega/\omega_n}{1 - (\omega/\omega_n)^2} \end{array}\right\}. \tag{4.61}$$

Figure 4.16 shows the Bode diagram of $G_2(j\omega)$ plotted against the dimensionless angular frequency ω/ω_n. Recalling Eq. (4.61), the resonance angular frequency ω_p at maximum value M_p of gain $|G_2(j\omega)|$ can be obtained as the angular frequency at $d\{[1 - (\omega/\omega_n)^2]^2 + (2\zeta\omega/\omega_n)^2\}/d(\omega)^2 = 0$.

$$\left.\begin{array}{l} \omega_p = \omega_n\sqrt{1 - 2\zeta^2} \\[2mm] M_p = \dfrac{1}{2\zeta\sqrt{1 - \zeta^2}} \end{array}\right\}. \tag{4.62}$$

The maximum value M_p of $|G(j\omega_p)|$ is called a resonance peak, and it is utilized as an index to rate the stability of the standard second order lag system because M_p is a function of the damping coefficient ζ.

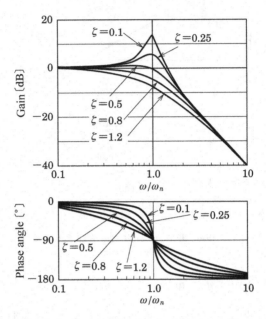

Fig. 4.16   Bode diagram of a standard second order system.

Dynamic performance is easily evaluated by the frequency response, with reference to a first or a second order lag system as follows:

(1) When the logarithmic gain $20 \log |G(j\omega)|$ of a static system deceases gradually with an increase in angular frequency ω for the low frequency range, and then it reaches $20 \log |G(j\omega)|_{\omega=0} - 3$ [dB] at angular frequency ω_b, which is called a bandwidth. The bandwidth is an index which may be used to evaluate the response speed. The static system without a resonance peak can be approximated as a first order lag system with equivalent time constant $T_e = 1/\omega_b$ or $T_e = 1/\omega_{bA}$, where ω_{bA} is the frequency at the phase angle $\angle G(j\omega_{bA}) = -45°$.

(2) Although the gain Bode diagram for an under-damped static system nearly maintains a constant value in the low frequency range, it gradually increases to the resonance peak and then rapidly decreases. The static system with a resonance peak is often approximated by a second order lag system whose undamped natural angular frequency ω_n and damping coefficient ζ are estimated making use of the resonance peak M_p and the resonance angular frequency ω_p in Eq. (4.62).

Table 4.7   Relations between the indicial response and frequency responses in a standard second order lag system.

	Indicial response	Frequency response
Equivalent time constant T_e	$T_e \approx \dfrac{2\zeta}{\omega_n}$ $(\zeta \geq 1)$	$T_e \approx \dfrac{1}{\omega_b}$
Maximum overshoot a_p	$a_p = \exp\left(-\dfrac{\pi\zeta}{\sqrt{1-\zeta^2}}\right)$ $(\zeta < 1)$	$a_p = \exp\left[\pi\left(\sqrt{M_p^2 - 1} - M_p\right)\right]$
Natural angular frequency ω_i	$\omega_i = \omega_n\sqrt{1-\zeta^2}$ $(\zeta < 1)$	$\omega_i = \omega_n\sqrt{\dfrac{1 + (\omega_p/\omega_n)^2}{2}}$

- ζ: Damping coefficient
- ω_n: Undamped natural angular frequency or approximated value by angular frequency at phase angle $-90°$
- ω_b: Bandwidth that is angular frequency at gain attenuation -3 dB or at $-45°$ of phase angle
- M_p: Resonance peak given by Eq. (4.62)
- ω_p: Resonance angular frequency given by Eq. (4.62)

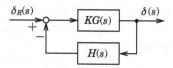

Fig. 4.17   Servo system.

(3) For a standard second order lag system, the relationships between the indicial response and the frequency response are summarized in Table 4.7.

4.2.5 *Performance of a Control Systems*

4.2.5.1 *Nyquist Stability Criterion*

Figure 4.17 shows a rearranged block diagram of a servo system whose closed loop transfer function $W(s)$ is defined as

$$W(s) = \frac{\delta(s)}{\delta_R(s)} = \frac{KG(s)}{1 + KG(s)H(s)}.$$ (4.63)

As shown in Fig. 4.18, both gain $|K\overline{G}(j\omega)|$ and phase angle $\angle K\overline{G}(j\omega)$ decrease with an increase of angular frequency ω, since the open loop transfer function $K\overline{G}(s) = KG(s)H(s)$ is a composite of the static systems

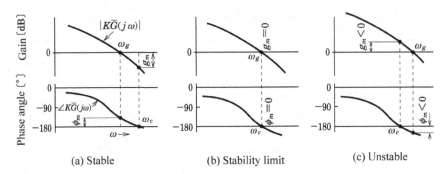

Fig. 4.18   Gain margin g_m and a phase margin ϕ_m.

transfer function and the integral elements, given by Eq. (4.44). Then, Nyquist stability criterion gives the following stability conditions.

(1) A feedback control system is stable for the phase angle $\angle K\overline{G}(j\omega_g) >$ $-180°$ on condition that the logarithmic gain $20\log|K\overline{G}(j\omega_g)|$ at angular frequency ω_g is zero.

(2) A feedback control system is stable for the logarithmic gain $20\log|K\overline{G}(j\omega_c)| < 0$ on condition that the phase angle $\angle K\overline{G}(j\omega_c)$ at frequency ω_c becomes $-180°$.

The angular frequency ω_g at logarithmic gain $20\log|K\overline{G}(j\omega_g)| = 0$ is referred to as a gain crossover frequency, and angular frequency ω_c at phase angle $\angle K\overline{G}(j\omega_c) = -180°$ is called a phase crossover frequency. The indicial response of a stable control system converges onto the commanded output value or a value with a steady error, whereas an unstable system, even with no input applied, will diverge due to an infinitesimal disturbance. For the stability limit in the phase angle $\angle K\overline{G}(j\omega_g) = -180°$ at the gain crossover frequency ω_g, the control system falls into the limit-cycle condition oscillating around the commanded value.

Thereupon, the magnitude of $\phi_m = \angle K\overline{G}(j\omega_g) + 180° > 0$ is called the phase margin, which may be used to rate the stability of a closed loop system. As an alternative to the phase margin, the gain margin $g_m = -20\log|K\overline{G}(j\omega_c)| > 0$ may also be used as a measure to rate the stability of a closed loop system.

$$\phi_m = \angle K\overline{G}(j\omega_g) + 180 \ [°], \tag{4.64}$$

$$g_m = -20\log|K\overline{G}(j\omega_c)| \ [\text{dB}]. \tag{4.65}$$

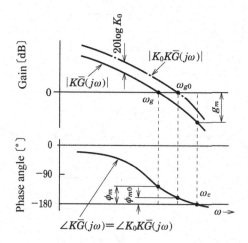

Fig. 4.19 Influence of the open loop gain K on the phase and gain margin.

The phase margin ϕ_m from 40 to 60° or the gain margin g_m from 10 to 20 dB should be at least given in the synthesis of an ordinary servo system. When the open loop frequency characteristics reveal a phase margin $\phi_m = 75°$ or more with the gain slope nearly 20 dB/decade around the gain crossover frequency, the control system may be approximated by the first order lag system with an equivalent constant $T_e = 1/\omega_b$ determined by the band width ω_b.

Figure 4.19 illustrates the Bode diagram of $K\overline{G}(s)$ and the Bode diagram of $K_0 K\overline{G}(s)$. The gain Bode diagram of $K_0 K\overline{G}(s)$ is obtained by shifting one of $K\overline{G}(s)$ as far as $20 \log K_0$ dB in vertical direction in parallel since the phase angle $\angle K_0 K\overline{G}(j\omega)$ coincides with the phase angle $\angle K\overline{G}(j\omega)$. The increase of the open loop gain from K to $K_0 K$ degrades the stability of the control system because the phase margin ϕ_m decreases to ϕ_{m0}, as shown in Fig. 4.19.

Figure 4.20 shows the block diagram of a control system whose open loop transfer function $K\overline{G}(s)$ is given by Eq. (4.66)

$$K\overline{G}(s) = \frac{\omega_n^2}{s(s + 2\zeta\omega_n)}. \tag{4.66}$$

Thus, the closed loop transfer function $C(s)/R(s)$ is represented by a standard second order lag system with the unit gain constant. The gain and the

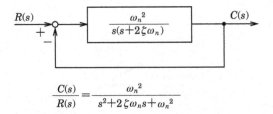

$$\frac{C(s)}{R(s)} = \frac{\omega_n^2}{s^2 + 2\zeta\omega_n s + \omega_n^2}$$

Fig. 4.20   Control system described by a standard second order lag system.

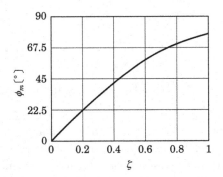

Fig. 4.21   Relationship between the phase margin and the damping coefficient.

phase angle of $K\overline{G}(j\omega)$ at the gain crossover frequency ω_g are expressed as

$$|K\overline{G}(j\omega_g)| = \frac{\omega_n^2}{\sqrt{\omega_g^4 + (2\zeta\omega_n\omega_g)^2}} = 1, \qquad (4.67)$$

$$\angle K\overline{G}(j\omega_g) = \tan^{-1}\frac{-2\zeta\omega_n}{-\omega_g} = -180° + \tan^{-1}\frac{2\zeta\omega_n}{\omega_g}. \qquad (4.68)$$

Solving Eq. (4.67), we get the gain crossover frequency ω_g

$$\omega_g = \omega_n\sqrt{\sqrt{1 + 4\zeta^4} - 2\zeta^2}. \qquad (4.69)$$

Recalling Eqs. (4.64), (4.68) and (4.69), the phase margin is expressed as

$$\phi_m = \tan^{-1}\frac{2\zeta\omega_n}{\omega_g} = \tan^{-1}\frac{2\zeta}{\sqrt{\sqrt{1 + 4\zeta^4} - 2\zeta^2}}. \qquad (4.70)$$

Figure 4.21 shows the relationship between the damping coefficients ζ and phase margins ϕ_m obtained from Eq. (4.70).

4.2.5.2 *Estimation of the Dynamic Performance by Means of a Root Locus Diagram*

As explained in Section 4.2.2, when the real parts of all the characteristic roots take minus values, the system is stable. Then, the few characteristic roots with a small absolute value of their real part may be regarded as the roots dominating the dynamic behavior throughout the whole dynamic performance. The characteristic roots dominating the dynamic performances are called dominant roots. The locations of the dominant roots on the complex plane give a criterion for evaluating the dynamic performance. For instance, when the system has only one evident dominant root $s = -1/T_e$ on the real axis, the system response may be approximated by a first order lag system with an equivalent time constant T_e.

Let us now consider a system with two evident dominant characteristic roots $-s_1$ and $-s_2$ defined by the following equation

$$-s_1 = -\zeta\omega_n + j\omega_n\sqrt{1-\zeta^2}, \quad -s_2 = -\zeta\omega_n - j\omega_n\sqrt{1-\zeta^2}. \quad (4.71)$$

Denoting the characteristic root $-s_1$ as a vector on a complex plane as shown in Fig. 4.22, the vector magnitude $|-s_1|$ is given by ω_n whilst the vector angle μ against the imaginary axis is defined as $\mu = \sin^{-1}\zeta$

$$\omega_n = |-s_1|, \quad \zeta = \sin\mu. \quad (4.72)$$

Since conjugate complex roots $-s_1$ and $-s_2$ correspond to the characteristic roots of a standard second order lag system given by Eq. (4.27), the system transfer function may be approximated by the standard second order lag system, defined by Eq. (4.26). The undamped natural angular frequency ω_n and damping coefficient ζ of the system can be estimated by the locations of the dominant roots on the complex plane, as shown in Fig. 4.22. The loop gain K in the control system shown in Fig. 4.17 affects

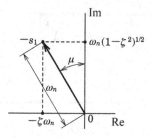

Fig. 4.22   Characteristic root $-s_1$ on a complex plane.

the location of the characteristic roots because the characteristic equation is given as

$$1 + K\overline{G}(s) = 0, \tag{4.73}$$

where $K\overline{G}(s) = KG(s)H(s)$. The characteristic roots reveal the loci on the complex plane associated with the variability of the loop gain K from zero to infinity, and the root loci with a graduated loop gain K is called a root locus diagram of the system. The root loci can be obtained by applying computer methods to solve the characteristic equation in which the loop gain K is varied systematically. This allows us to select the optimal loop gain when synthesizing the control system. For instance, let us consider a system whose open loop transfer function $K\overline{G}(s)$ is given as

$$K\overline{G}(s) = \frac{K}{s(s+2)(s+4)}. \tag{4.74}$$

The characteristic equation of the control system is written as

$$s^3 + 6s^2 + 8s + K = 0. \tag{4.75}$$

Figure 4.23 shows the root locus diagram obtained by solving Eq. (4.75) using computer software. Equation (4.75) for $K = 0$ gives characteristic roots, $-s_1 = 0$, $-s_2 = -2$ and $-s_3 = -4$ which are plotted on the real

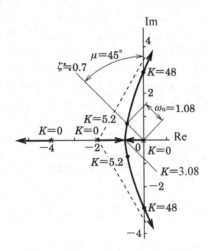

Fig. 4.23   Root locus diagram of Eq. (4.75).

axis using the symbols ×. Each characteristic root moves in the direction of the arrow as the loop gain K increases. The characteristic roots $-s_1$ and $-s_2$ merge at $K = 3.08$ and become the conjugate complex roots for $K > 3.08$, and then extend further to infinity for $K \to \infty$, whereas the other characteristic root $-s_3$ is located further on the real axis. The root locus diagram implies that the system performance at $K = 5.2$ can be approximated by a standard second order lag system with $\zeta = 0.7$ and $\omega_n = 1.08\,\mathrm{rad/s}$ because the dominant roots at $K = 5.2$ are located on the straight line defined $\zeta = \sin 45° \approx 0.7$. The stability of the control system degrades as the loop gain K increases, and then reaches the stability limit on the imaginary axis at $K = 48$. The system becomes unstable for $K > 48$ because the real parts of the dominant roots take the plus values.

Let us consider general rules governing the root loci. Though the root locus diagram is typically obtained by solving the characteristic equation, the rules governing root loci developed by W. R. Evans are useful for grasping the image of the whole root locus diagram. This is because the rules give a schematic concept of the root locus diagram for the loop gain K from zero to infinity. Let us now consider a control system with the open loop transfer function $K\overline{G}(s)$ written in the factorized form

$$K\overline{G}(s) = \frac{K_0(s + z_1)(s + z_2)\cdots(s + z_i)\cdots(s + z_m)}{(s + p_1)(s + p_2)\cdots(s + p_k)\cdots(s + p_n)}. \tag{4.76}$$

In the transfer function denoted by Eq. (4.76), the values of $s = -p_1, -p_2 \cdots, -p_n$ and $s = -z_1, -z_2 \cdots, -z_n$ are referred as poles and zeros, respectively.

Recalling Eq. (4.73), the complex variables s on the root loci satisfy the following characteristic equation:

$$\frac{K_0(s + z_1)(s + z_2)\cdots(s + z_i)\cdots(s + z_m)}{(s + p_1)(s + p_2)\cdots(s + p_k)\cdots(s + p_n)} = -1. \tag{4.77}$$

Since the arbitrary point s on the root loci is represented by a vector $\vec{s}$ as shown in Fig. 4.24, the terms $s + z_i$ and $s + p_k$ in the transfer function $K\overline{G}(s)$ are also represented by vectors directed towards the location of complex variables s on the loot loci from zero $-z_i$ and from pole $-p_k$, respectively.

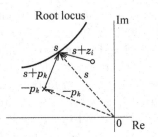

Fig. 4.24   Vectors $\vec{a}$, $\overrightarrow{s+p_k}$ and $\overrightarrow{s+z_i}$.

Recalling mathematical formula $e^{j\theta} = \cos\theta + j\sin\theta$, the terms $s + p_k$ and $s + z_i$ are expressed as

$$\left.\begin{array}{c} s + p_k = A_k e^{j\alpha_k} \\ s + z_i = B_i e^{j\beta_k} \end{array}\right\}, \tag{4.78}$$

where A_k and α_k denote the magnitude and the angle of vector $\overrightarrow{s+p_k}$, and B_i and β_i are the magnitude and the angle of vector $\overrightarrow{s+z_i}$, respectively. Substituting Eq. (4.78) into Eq. (4.77) gives

$$\left.\begin{array}{c} K\overline{G}(s) = \left(\dfrac{K_0 \prod\limits_{i=1}^{m} B_i}{\prod\limits_{k=1}^{n} A_k}\right) \exp j \left(\sum\limits_{i=1}^{m} \beta_i - \sum\limits_{k=1}^{n} \alpha_k\right) = -1 \\[2em] \prod\limits_{k=1}^{n} A_k = A_1 A_2 \cdots A_n, \quad \prod\limits_{i=1}^{m} B_i = B_1 B_2 \cdots B_m \end{array}\right\}. \tag{4.79}$$

Equation (4.79) yields Eqs. (4.80) and (4.81) which give the variable s on the root loci

$$|K\overline{G}(s)| = \frac{K_0 \prod\limits_{i=1}^{m} B_i}{\prod\limits_{k=1}^{n} A_k} = 1. \tag{4.80}$$

$$\angle K\overline{G}(s) = \sum\limits_{i=1}^{m} \beta_i - \sum\limits_{k=1}^{n} \alpha_k = \pm(2N_i + 1)\pi \quad (N_i = 0, 1, 2, \ldots). \tag{4.81}$$

Equations (4.80) and (4.81) are called a gain condition and a phase angle condition, respectively. Equations (4.77) to (4.81) underpin the following rules governing the loot loci.

(1) A root locus diagram is a symmetrical about the real axis.

(2) A root locus starts at $K = 0$ and the starting points are the poles of the open loop transfer function. The root locus terminal at $K \to \infty$ approaches to the position of the zeros in the open loop transfer function or infinitely far points. Therefore, denoting the number of zeros and poles by m and n respectively, the number of branches tending to infinity becomes $n - m$.

(3) When the total number of poles and zeros located on the right-hand side of the real axis toward an arbitrary point on the real axis is odd, the region on the right side axis belongs to the root locus branch. For instance, Fig. 4.23 shows that a line on the real axis for $-s_3 < -4$ is a root locus branch because there are three poles on the right-hand side of the real axis toward an arbitrary point on the real axis for $-s_3 < -4$.

(4) When the asymptote of a root locus intersects the real axis at the location C inclined at the angle Θ, the inclination angle Θ and location C are expressed as

$$C = \frac{\sum\limits_{k=1}^{n} (-p_k) - \sum\limits_{i=1}^{m} (-z_i)}{n - m}, \qquad (4.82)$$

$$\Theta = \frac{(2N_i + 1)\pi}{n - m}, \qquad (4.83)$$

where N_i is integer $(N_i = 0, 1, 2, 3, \ldots)$.

(5) When two branches of the root loci on the real axis merge at a point by increasing the loop gain, then they break away from the real axis. Two merged characteristic roots $s_1 = s_2 = \sigma$ at the break away point can be derived from the following formula:

$$\sum_{k=1}^{n} \frac{1}{\sigma + p_k} = \sum_{i=1}^{m} \frac{1}{\sigma + z_i}. \qquad (4.84)$$

4.2.5.3 *Hurwitz Stability Criterion*

The characteristic equation of a system is written as

$$a_0 s^n + a_1 s^{n-1} + a_2 s^{n-2} + \cdots + a_{n-1} s + a_n = 0. \qquad (4.85)$$

The Hurwitz stability criterion imposes the conditions on the coefficients $a_0, a_1, a_2, \ldots, a_n$ for the system to be stable. The stability conditions are defined as follows.

Condition 1: All coefficients $a_0, a_1, a_2, \ldots, a_n$ are positive.

Condition 2: Determinants $\Delta_n, \Delta_{n-1}, \cdots \Delta_3, \Delta_2$ given by following Eq. (4.86) are all positive.

$$
\left.
\begin{aligned}
\Delta_n &=
\begin{vmatrix}
a_1 & a_3 & a_5 & a_7 & \cdots & \cdots & \cdots & \cdots & \cdots & 0 \\
a_0 & a_2 & a_4 & a_6 & \cdots & \cdots & \cdots & \cdots & \cdots & 0 \\
0 & a_1 & a_3 & a_5 & \cdots & \cdots & \cdots & \cdots & \cdots & 0 \\
0 & a_0 & a_2 & a_4 & \cdots & \cdots & \cdots & \cdots & \cdots & 0 \\
\cdots & \cdots & \cdots & \cdots & \cdots & \cdots & \cdots & \cdots & \cdots & \cdots \\
0 & 0 & 0 & 0 & \cdots & \cdots & \cdots & \cdots & \cdots & a_n
\end{vmatrix} > 0 \\[2mm]
\Delta_2 &=
\begin{vmatrix}
a_1 & a_3 \\
a_0 & a_2
\end{vmatrix} > 0, \quad
\Delta_3 =
\begin{vmatrix}
a_1 & a_3 & a_5 \\
a_0 & a_2 & a_4 \\
0 & a_1 & a_3
\end{vmatrix} > 0, \cdots \quad \cdots, \quad \Delta_{n-1} > 0
\end{aligned}
\right\}.
$$

$$(4.86)$$

Let us now consider the characteristic equation given by Eq. (4.75). Though the condition 1 is satisfied for the loop gain $K > 0$, the following determinants involved in condition 2 must be also satisfied for the system to be stable

$$
\Delta_3 =
\begin{vmatrix}
6 & K & 0 \\
1 & 8 & 0 \\
0 & 6 & K
\end{vmatrix} = K(48 - K) > 0, \quad
\Delta_2 =
\begin{vmatrix}
6 & K \\
1 & 8
\end{vmatrix} = 48 - K > 0.
$$

$$(4.87)$$

Therefore, the system is stable for $0 < K < 48$, and it falls below the stability limit at $K = 48$. The stability limit at $K = 48$ is also confirmed by the root locus diagram shown in Figure 4.23, because the two root loci branches intersect the imaginary axis at $K = 48$.

4.2.6 *Method of Identifying the Transfer Function of Hydraulic Components*

Though the performance of a hydraulic component is essentially nonlinear, it can often be approximated by a standard second order lag system. However, it is a formidable task to identify the approximated transfer function of the hydraulic components by measuring the frequency responses. This is because it varies with the load condition, magnitude of sinusoidal input

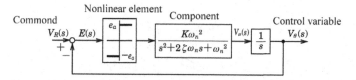

Fig. 4.25 Self-excited oscillation system.

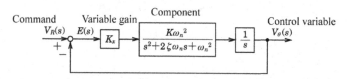

Fig. 4.26 Control system replacing an on–off element with a variable gain.

and oil temperature, etc. This section presents a novel method that can be effectively employed to identify the transfer function of the hydraulic components without measuring the frequency responses.

Figure 4.25 shows a self-excited oscillation system comprising an on–off element, a hydraulic component to identify the approximated second order lag system and an integration element. The on–off element generates the voltage e_a or $-e_a$, depending on the plus or minus sign of the error signal voltage $e = v_R - v_\theta$. Therefore, the on–off element may be regarded as a variable gain $K_x = e_a/|e|$ varying with the error signal. Variable gain $K_x = e_a/|e|$ increases excessively when the control variable v_θ approaches the command value v_R, and the system becomes unstable and the error magnitude $|e|$ increases. Since the increase of the error magnitude $|e|$ makes the system stable due to the decrease of variable gain $K_x = e_a/|e|$, the control variable will again approach the command value. Consequently, the control variable oscillates around the command value with an amplitude that is only a little larger than the stability limit error e_c. The stability limit error e_c stands for the error at the stability limit gain $K_{xc} = e_a/|e_c|$.

Figure 4.26 shows the control system wherein the on–off element in Fig. 4.25 is replaced by a variable gain K_x. The characteristic equation is given as

$$s^3 + 2\zeta\omega_n s^2 + \omega_n^2 s + K_x K \omega_n^2 = 0. \qquad (4.88)$$

The Hurwitz stability criterion gives the stability limit gain K_{xc} as follows:

$$K_{xc} = \frac{e_a}{e_c} = \frac{2\zeta\omega_n}{K}. \qquad (4.89)$$

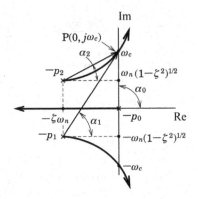

Fig. 4.27   Root locus diagram and vector angles α_1, α_2 and α_3.

Figure 4.27 shows the root locus diagram of the system with the characteristic equation given by Eq. (4.88). The root locus starting from the pole $-p_0 = 0$ moves along the real axis in the minus infinity direction and the root loci starting from the other poles $-p_1, -p_2 = -\zeta\omega_n \pm j\omega_n\sqrt{1-\zeta^2}$ pass through the stability limit points $\pm j\omega_c$ respectively, and then tend towards infinity. Applying Eq. (4.81) to point $s = j\omega_c$ on the root locus gives $\alpha_0 + \alpha_1 + \alpha_2 = \pi$ in Fig. 4.27.

Therefore,

$$\tan^{-1}\frac{\omega_c - \omega_n\sqrt{1-\zeta^2}}{\zeta\omega_n} + \tan^{-1}\frac{\omega_c + \omega_n\sqrt{1-\zeta^2}}{\zeta\omega_n} = \frac{\pi}{2}. \qquad (4.90)$$

By virtue of the trigonometric function formula $\tan^{-1}\alpha_1 + \tan^{-1}\alpha_2 = \tan^{-1}[(\alpha_1 + \alpha_2)/(1 - \alpha_1\alpha_2)]$, Eq. (4.90) is rearranged as follows:

$$\tan^{-1}\frac{2\zeta\omega_c\omega_n}{\omega_n^2 - \omega_c^2} = \frac{\pi}{2}. \qquad (4.91)$$

The angular frequency ω_c satisfying Eq. (4.91) is given as follows:

$$\omega_c = \omega_n. \qquad (4.92)$$

In the light of the oscillatory behavior produced in the self-excited oscillation system due to the fluctuation of gain K_x around stability limit gain $K_{xc} = e_a/e_c$, it supposed that the angular frequency ω_s of the self-exited oscillation system is aligned to the angular frequency $\omega_c = \omega_n$ given by Eq. (4.92). Furthermore, the amplitude V_A of self-excited oscillation is significantly affected by the magnitude of the stability limit error e_c. Let the self-excited oscillation performance be expressed in terms of angular frequency ω_c and stability limit error e_c

$$\xi = \frac{\omega_s}{\omega_n}, \tag{4.93}$$

$$R = \frac{V_A}{e_c}, \tag{4.94}$$

where ξ and R are the correction factors. Recalling Eqs. (4.89), (4.92), (4.93) and (4.94), dynamic parameters ω_n and ζ of the hydraulic component are given as

$$\omega_n = \frac{\omega_s}{\xi}, \tag{4.95}$$

$$\zeta = \frac{RKe_a}{2V_A\left(\omega_s/\xi\right)}. \tag{4.96}$$

Theoretical considerations imply that the correction factor ξ is a function of the damping coefficient ζ, and that the relationship between ξ and R may be denoted as an implicit function. The approximated transfer function being identified, Eqs. (4.97) and (4.98) obtained by computer simulations will be used for convenience. For $0.1 \leq \zeta \leq 10$

$$R = 1.27 + 0.0647\zeta - 0.00762\zeta^2 + 0.000307\zeta^3, \tag{4.97}$$

$$\xi = 1.0 + 0.0315\zeta + 0.00415\zeta^2 - 0.000185\zeta^3. \tag{4.98}$$

Although the correction factors ξ and R depend on the damping coefficient ζ, the changes in ξ and R are relatively small in comparison with variations of ζ. Therefore, the damping coefficient ζ and the undamped natural frequency ω_n of a hydraulic component can be identified by the following procedure which is called the self excited oscillation method (See Refs. [10] and [13] in detail).

(1) Configure a self-excited oscillation system including a hydraulic component to identify the approximated second order lag transfer function, noting that the amplitude voltage of V_A and angular frequency ω_s in the self-excited oscillation system must be measured at first.

(2) Estimate the values of $\xi = \xi_1$ and $R = R_1$ for an assumed value of the damping coefficient $\zeta = \zeta_{a1}$ recalling Eqs. (4.97) and (4.98) respectively, and then estimate the damping coefficient $\zeta = \zeta_1$ by substituting $\xi = \xi_1$

and $R = R_1$, etc. into Eq. (4.96). If the assumed damping coefficient $\zeta = \zeta_{a1}$ is smaller (larger) than the true damping coefficient, the estimated damping coefficient $\zeta = \zeta_1$ is larger (smaller) than the true damping coefficient because the increase rate of R for the change of ζ exceeds the decrease rate of ξ.

(3) Adopt a more appropriate value of the damping coefficient $\zeta_{a2} = (\zeta_{a1} + \zeta_1)/2$ and then estimate $R = R_2$, $\xi = \xi_2$ recalling Eqs. (4.97) and (4.98), respectively. Substitute the values $\xi = \xi_2$ and $R = R_2$, etc. into Eq. (4.96) and compute the damping coefficient $\zeta = \zeta_2$ accordingly. Iterate the same procedure k-times so that the estimated damping coefficient $\zeta = \zeta_k$ should correspond to the true damping coefficient of the hydraulic component. The undamped natural frequency ω_n of the hydraulic component may be obtained from Eq. (4.95) because the true value of the parameter $\xi = \xi_k$ for $\zeta = \zeta_k$ may be derived from Eq. (4.98).

4.3 Dynamic Characteristics of Hydraulic Control Components

4.3.1 *Transducers and Servo Amplifiers*

A transducer transforms a physical quantity into an electrical voltage signal in order to deliver it in a control system. Linear/rotary potentiometers and linear/rotary variable differential transducers (LVDT/RVDT) are used as analogue type transducers that detect linear or rotary displacement. A potentiometer consists of a metalized ceramic electric resistance with a DC source and a slider on the resistance. The voltage across the potentiometer is proportional to the slider displacement. A variable differential transducer is a displacement detector that uses electromagnetic induction generated by the prime coil with an AC source. It produces a DC voltage in accordance with the core displacement in the secondary coil using an AD converter. Comparing a variable differential transducer with a potentiometer reveals a few obvious advantages of the former, such as frictionless operation, infinite resolution with the analogue signal and superior endurance under shocks. As a velocity transducer, a DC tacho-generator utilizing Faraday's right hand law can be used in a hydraulic control system. The response speeds in these transducers are so high that the relationship between the transducer output voltage $v(t)$ and input $x(t)$ can be given as $v(t) = K_t x(t)$, using the transformation constant K_t. Therefore, the transfer function is given as follows:

$$G_t(s) = \frac{V(s)}{X(s)} = K_t. \tag{4.99}$$

A servo amplifier comprises an operational network and a power amplifier which provides the electric current required to drive the servo valve. Since the operating time in a servo amplifier is about $0.5\,\text{ms}$ in general, it may be regarded as negligibly small compared to the response of an entire hydraulic control system. Consequently, the transfer function of the servo amplifier can be also expressed:

$$G_a(s) = \frac{I(s)}{V(s)} = K_a, \tag{4.100}$$

where $V(s)$ and $I(s)$ stand for the Laplace transform of the electrical voltage input $v(t)$ and the electric current output $i(t)$ out of the amplifier, respectively. Most servo amplifiers provide not only the adjustable constant K_a but also a compensation element to improve the dynamic performance of the control system. The transfer function of the amplifier with a phase lead or lag compensation element is given as

$$G_a(s) = \frac{I(s)}{V(s)} = K_a G_c(s) = K_a \frac{1 + aTs}{1 + Ts}. \tag{4.101}$$

The transfer function of an amplifier with a Proportion/Integration/Differentiation compensator (PID compensator) is given as follows:

$$G_a(s) = \frac{I(s)}{V(s)} = K_a G_c(s) = K_a K_p \left(1 + \frac{1}{T_i s} + T_d s \right), \tag{4.102}$$

where the parameters K_a, a, T in Eq. (4.101) and K_a, K_p, T_i, T_d in Eq. (4.102) are adjustable constants that should be selected so as to achieve the most desirable dynamic characteristics of the control system. The method employed to select the desired values of these parameters will be discussed in Section 4.4.3.

Figure 4.28 illustrates an operational amplifier called a differential amplifier. When two input voltages e_1, e_2 are applied to an operational amplifier, it produces an output voltage of $e_o = \mu(e_1 - e_2)$. The differential amplifier gain μ is in the range from 10^5 to 10^6, and the electric input impedance is so high that the electric current hardly flows into the operational amplifier. Moreover, the electric output impedance is so low that

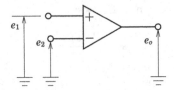

Fig. 4.28   Differential amplifier.

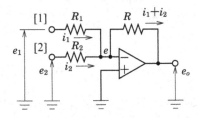

Fig. 4.29   Additional network.

the output voltage is hardly affected by the load connected to the output terminal

$$e_o = \mu(e_1 - e_2). \qquad (4.103)$$

Figure 4.29 illustrates an additional network comprising a differential amplifier and the impedances. Applying input voltages e_1 and e_2 to the input terminals [1] and [2], respectively, the relationships between the impedance and the voltage in the additional network are described as follows:

$$\left.\begin{aligned}
e_1 - e &= R_1 i_1 \\
e_2 - e &= R_2 i_2 \\
e - e_o &= R(i_1 + i_2) \\
-\mu e &= e_o
\end{aligned}\right\}, \qquad (4.104)$$

where i_1 and i_2 are the electric currents flowing through impedances R_1 and R_2 respectively, and then electrical current $i_1 + i_2$ should pass through the impedance R in the additional network. Cancelling variables i_1, i_2 and e in Eq. (4.104), the output voltage e_0 from the additional network becomes

$$e_o = -\left(\frac{R}{R_1}e_1 + \frac{R}{R_2}e_2\right). \qquad (4.105)$$

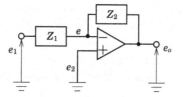

Fig. 4.30 Operational network.

Table 4.8 Impedance Z_1 and Z_2 for various transfer function circuits.

	Impedance Z_1	Impedance Z_2
(a) Addition $$E_0(s) = -\left[\frac{R}{R_1}E_1(s) + \frac{R}{R_2}E_2(s)\right]$$	$e_1 \circ\!-\!\!\text{W}\!\!\!\!-\!\bullet\!-\!\circ\ e$ R_1 $e_2 \circ\!-\!\!\text{W}\!\!\!\!-$ R_2	$e \circ\!-\!\!\text{W}\!\!\!\!-\!\circ\ e_o$ R
(b) Integral $$\frac{E_0(s)}{E_1(s)} = -\frac{1}{CR_1 s}$$	e_1 R_1 e $\circ\!-\!\!\text{W}\!\!\!\!-\!\circ$	e C e_o $\circ\!-\!\|\!-\!\circ$
(c) Proportional and derivative $$\frac{E_0(s)}{E_1(s)} = -\frac{R}{R_1}(1 + C_1 R_1 s)$$	e_1 R_1 e $\circ\!-\!\!\text{W}\!\!\!\!-\!\circ$ $\|\!-$ C_1	e R e_o $\circ\!-\!\!\text{W}\!\!\!\!-\!\circ$
(d) Phase lag compensation $$\frac{E_0(s)}{E_1(s)} = -\left(\frac{R}{R_1}\right)\frac{1+aTs}{1+Ts}$$ $$a = \frac{R_0}{R_0+R} < 1, \quad T = (R_0+R)C$$	e_1 R_1 e $\circ\!-\!\!\text{W}\!\!\!\!-\!\circ$	e R e_o $\circ\!-\!\!\text{W}\!\!\!\!-\!\circ$ $\|\!-\!\text{W}$ C R_o
(e) Phase lead compensation $$\frac{E_0(s)}{E_1(s)} = -\left(\frac{R}{R_1}\right)\frac{1+aTs}{1+Ts}$$ $$a = \frac{R_1+R_2}{R_2} > 1, \quad T = C_1 R_2$$	e_1 R_1 e $\circ\!-\!\!\text{W}\!\!\!\!-\!\circ$ $\|\!-\!\text{W}$ C_1 R_2	e R e_o $\circ\!-\!\!\text{W}\!\!\!\!-\!\circ$

Figure 4.30 illustrates a most general form of an operational network whose impedances Z_1 and Z_2 provide the various transfer functions as shown in Table 4.8.

The following procedure is the recommended way to derive the transfer function of an operational network.

(1) In Fig. 4.30, input voltage e at the minus feedback circuit can be treated as zero because the magnitude of the differential amplifier gain μ can be regarded as infinite when compared with the finite output voltage e_0.

(2) Electric current flows only through impedances Z_1 and Z_2 because the impedance of an operational amplifier may be regarded as infinite.

For instance, substituting $e = 0$ into Eq. (4.104) gives

$$\left.\begin{aligned}
e_1 &= R_1 i_1 \\
e_2 &= R_2 i_2 \\
-e_0 &= R(i_1 + i_2)
\end{aligned}\right\}. \tag{4.106}$$

Canceling currents i_1 and i_2 in Eq. (4.106) gives Eq. (4.105).

4.3.2 Servo valves

4.3.2.1 Transfer Function of a Servo valve

A servo valve controls the flow direction and flow rate into the actuator, depending on the output signal of the servo amplifier. As explained in Section 3.3.7, the nozzle flapper in the servo valve inclines at a small angle θ which is proportional to the electric current i flowing into the torque motor coil in the servo valve. The flapper response speed is so high in comparison with the spool response that the transfer function $\theta(s)/I(s)$ may be defined as

$$\frac{\theta(s)}{I(s)} = K_\theta. \tag{4.107}$$

Figure 4.31 illustrates the dimensional specification of the servo valve. The nozzle diameter and the distance between the nozzle and the flapper being denoted by d_n and z_n, respectively, the variable opening nozzle port area A_1 is expressed as follows:

$$A_1 = \pi d_n z_n. \tag{4.108}$$

Thus, the flow rate Q_1 out of the nozzle is described as

$$Q_1 = \pi C_{d1} d_n z_n \sqrt{\frac{2p_n}{\rho}}, \tag{4.109}$$

where C_{d1}, p_n and ρ denote the flow rate coefficient, the pressure in the nozzle chamber and the fluid density, respectively. The flow rate Q_2 through

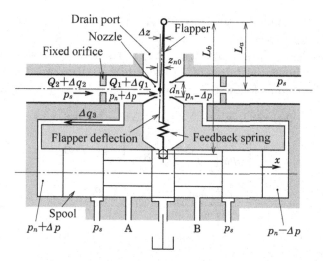

Fig. 4.31   Servo valve (Force feedback type).

the fixed orifice with area A_2 is expressed as follows:

$$Q_2 = C_{d2} A_2 \sqrt{\frac{2(p_s - p_n)}{\rho}}, \tag{4.110}$$

where p_s and C_{d2} denote the pressure upstream of the orifice and the flow rate coefficient, respectively.

As implied by Eq. (4.109) and (4.110), the flow rate Q_1 is a nonlinear function on variables p_n and z_n, and Q_2 is the nonlinear function on the variable p_n. However, these equations may be linearized as long as the distance z_n between the flapper and the nozzle changes infinitesimally from the neutral position $z_n = z_{n0}$ to $z_n = z_{n0} - \Delta z$.

Let us now consider the procedure for linearization of Eq. (4.109). When the distance between the flapper and nozzle changes infinitesimally from $z_n = z_{n0}$ to $z_n = z_{n0} - \Delta z$, the left nozzle opening area A_1 and pressure p_n in the left nozzle chamber change from $A_1 = \pi d_n z_{n0}$ to $A_1 = \pi d_n(z_{n0} - \Delta z)$ and from $p_n = p_{n0}$ to $p_n = p_{n0} + \Delta p$ respectively, whereas the right nozzle opening area A_{1R} and pressure p_{nR} in the right nozzle chamber also change from $A_{1R} = \pi d_n z_{n0}$ to $A_{1R} = \pi d_n(z_{n0} + \Delta z)$ and from $p_{nR} = p_{n0}$ to $p_n = p_{n0} - \Delta p$, respectively. The hydraulic force $2\Delta p A_s$ acts on the spool with the cross-section area A_s and the spool moves to the right. Consequently, the flow rate Q_1 given by Eq. (4.109) changes from $Q_1(p_{n0}, z_{n0})$ to $Q_1(p_{n0} +$

$\Delta p, z_{n0} - \Delta z$), and the flow rate Q_2 given by Eq. (4.110) changes from $Q_2(p_{n0})$ to $Q_2(p_{n0} + \Delta p)$.

Developing $Q_1(p_{n0} + \Delta p, z_{n0} - \Delta z)$ in the form of Taylor series and neglecting exponential terms of an order higher than two, the flow rate change is derived as follows:

$$\Delta q_1 = Q_1(p_{n0} + \Delta p, z_{n0} - \Delta z) - Q_1(p_{n0}, z_{n0})$$

$$= \left[\frac{\partial Q_1}{\partial p_n}\right]_{\substack{z_{n0} \\ p_{n0}}} \Delta p + \left[\frac{\partial Q_1}{\partial z_n}\right]_{\substack{z_{n0} \\ p_{n0}}} (-\Delta z)$$

$$= \frac{\pi C_{d1} d_n z_{n0}}{\sqrt{2\rho p_{n0}}} \Delta p - \pi C_{d1} d_n \sqrt{\frac{2p_{n0}}{\rho}} \Delta z. \tag{4.111}$$

The flow rate change $\Delta q_2 = Q_2(p_{n0} + \Delta p) - Q_2(p_{n0})$ can be obtained in the similar manner.

$$\Delta q_2 = Q_2(p_{n0} + \Delta p) - Q_2(p_{n0}) = \left[\frac{dQ_2}{dp_n}\right]_{p_{n0}} \Delta p = -\frac{C_{d2} A_2}{\sqrt{2\rho(p_s - p_{n0})}} \Delta p, \tag{4.112}$$

where suffixes p_{n0}, z_{n0} denote the differential value at $p_n = p_{n0}$, $z_n = z_{n0}$.

Denoting the spool cross-section area and the spool displacement for the infinitesimal time dt by A_s and Δx respectively, the hydraulic capacity $(\Delta q_2 - \Delta q_1)dt$ of fluid flowing into the spool chamber within time dt equals the spool displacement volume $A_s \Delta x$. Therefore, the flow rate $\Delta q_2 - \Delta q_1 = A_s(d\Delta x/dt)$ can be given as follows:

$$\left.\begin{aligned} A_s \frac{d\Delta x}{dt} &= -C_1 \Delta p + \pi C_{d1} d_n \sqrt{\frac{2p_{n0}}{\rho}} \Delta z \\ C_1 &= \frac{C_{d2} A_2}{\sqrt{2\rho(p_s - p_{n0})}} + \frac{\pi C_{d1} d_n z_{n0}}{\sqrt{2\rho p_{n0}}} \end{aligned}\right\}. \tag{4.113}$$

The equation of the spool motion is defined as follows:

$$2A_s \Delta p = m\frac{d^2 \Delta x}{dt^2} + B\frac{d\Delta x}{dt} + kL_b \Delta\theta, \tag{4.114}$$

where m is the spool mass, B is a coefficient of the viscous resistance force in the spool motion, k is a spring constant of the feedback spring, $\Delta\theta$ is the inclined angle of the flapper, L_b is the distance from the flapper fulcrum to the connection point with the spool, and L_a is the distance from the flapper fulcrum to the nozzle jet position, as shown in Fig. 4.31.

Though the torque motor supplies the flapper displacement $L_a\Delta\theta$ at the nozzle jet position at first because of the very small moment of inertia, the flapper displacement is returned back as far as $(L_a/L_b)\Delta x$ through the feedback spring because the hydraulic force $2\Delta pA_s$ between both spool ends gives the spool displacement Δx in the right direction. Consequently, the flapper torque produced by the torque motor is balanced by the spool force torque $2L_b\Delta pA_s$ through the feedback spring, which is in a steady state. Therefore, the net displacement Δz of the flapper at the nozzle jet position is given as

$$\Delta z = L_a\Delta\theta - \frac{L_a}{L_b}\Delta x. \tag{4.115}$$

Taking all of the initial conditions as zero, the Laplace transforms of Eqs. (4.113), (4.114), (4.115) are expressed as follows:

$$\begin{bmatrix} A_s s & C_1 & -\pi C_{d1}d_n\sqrt{2p_{n0}/\rho} \\ ms^2 + B_s & -2A_s & 0 \\ L_a/L_b & 0 & 1 \end{bmatrix} \begin{bmatrix} \Delta X(s) \\ \Delta P(s) \\ \Delta Z(s) \end{bmatrix} = \begin{bmatrix} 0 \\ -kL_b \\ L_a \end{bmatrix} \Delta\theta(s). \tag{4.116}$$

Canceling $\Delta P(s)$ and $\Delta Z(s)$ in Eq. (4.116) gives

$$\frac{\Delta X(s)}{\Delta\theta(s)} = \frac{\frac{2\pi C_{d1}L_a d_n A_s}{C_1}\sqrt{\frac{2p_{n0}}{\rho}} - kL_b}{ms^2 + \left(B + \frac{2A_s^2}{C_1}\right)s + \frac{2\pi C_{d1}d_n A_s L_a}{C_1 L_b}\sqrt{\frac{2p_{n0}}{\rho}}}. \tag{4.117}$$

Since the flow rate change Δq_{m0} through the servo valve is proportional to the spool displacement change Δx without load condition, the transfer function is described as

$$\frac{\Delta Q_{m0}(s)}{\Delta X(s)} = K_s, \tag{4.118}$$

where $\Delta X(s)$ and $\Delta Q_{m0}(s)$ are Laplace transforms of Δx and Δq_{m0}, respectively. Thus, Eq. (4.107) can be rearranged as

$$\frac{\Delta\theta(s)}{\Delta I(s)} = K_\theta. \tag{4.119}$$

Canceling $\Delta X(s)$ and $\Delta\theta(s)$ in Eqs. (4.117), (4.118) and (4.119) gives

$$G_v(s) = \frac{\Delta Q_{m0}(s)}{\Delta I(s)} = \frac{K_s K_\theta \left[\left(\frac{2\pi C_{d1}L_a d_n A_s}{C_1}\sqrt{\frac{2p_{n0}}{\rho}}\right) - kL_b\right]}{ms^2 + \left(B + \frac{2A_s^2}{C_1}\right)s + \frac{2\pi C_{d1}d_n A_s L_a}{C_1 L_b}\sqrt{\frac{2p_{n0}}{\rho}}}. \tag{4.120}$$

Consequently, the transfer function of servo valve flow rate change ΔQ_{m0} for an infinitesimal change in electric current ΔI may be rearranged in the form of a standard second order lag system:

$$G_v(s) = \frac{\Delta Q_{m0}(s)}{\Delta I(s)} = \frac{K_v \omega_{nv}^2}{s^2 + 2\zeta_v \omega_{nv} s + \omega_{nv}^2}, \qquad (4.121)$$

$$\left. \begin{aligned} 2\zeta_v \omega_{nv} &= \frac{1}{m}\left(B + \frac{2A_s^2}{C_1}\right) \\ \omega_{nv}^2 &= \frac{2\pi C_{d1} d_n A_s L_a}{m C_1 L_b}\sqrt{\frac{2p_{n0}}{\rho}} \\ K_v \omega_{nv}^2 &= \frac{K_s K_\theta}{m}\left(\frac{2\pi C_{d1} L_a d_n A_s}{C_1}\sqrt{\frac{2p_{n0}}{\rho}} - kL_b\right) \end{aligned} \right\}. \qquad (4.122)$$

4.3.2.2 *Dynamic Flow Characteristics of a Servo valve without Load Condition*

Let us assume that for a small change of state, the dynamic performance of a servo valve can be approximated by Eqs. (4.121), (4.122). However, the actual dynamic characteristic of a servo valve is both nonlinear and complex. This is due to leakage and friction between the spool and the spool bore. Moreover it is difficult in practice to make a theoretical approximation of the dynamic performance because the dimensional specifications such as $C_1, z_{n0}, L_a \cdots$, etc. are themselves not straightforward. As a result, the dynamic performance of a servo valve is commonly identified by measuring the amplitudes and the phase angles in the frequency responses for various frequencies ω and under specified conditions.

Figure 4.32 illustrates the frequency response method for identifying the dynamic performance of a servo valve. If the frequency voltage input $V = e_a \sin \omega t$ is supplied to the servo amplifier, then the voltage output $V_\omega = V_0 \sin(\omega t + \phi)$ corresponding to the output flow rate through the servo valve is obtained by measuring the speed of the cylinder piston with a negligibly small mass. Identification of the approximate transfer function by the frequency method is a troublesome task because it is necessary to measure the frequency responses for various conditions and to identify the approximated transfer functions on the basis of measurements of the frequency characteristics.

Once it is established that the hydraulic component performance can be approximated by the formula applicable to the standard second order

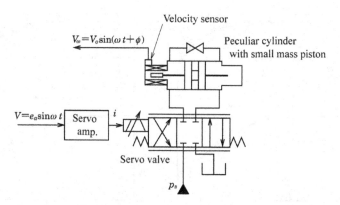

Fig. 4.32 Method to measure the frequency response of a servo valve.

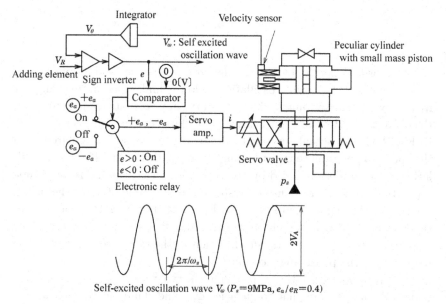

Self-excited oscillation wave V_ω (P_s=9MPa, e_a/e_R=0.4)

Fig. 4.33 Self-excited oscillation system to identify the servo valve transfer function.

lag system, then the approximate transfer function is easily identified using the self-excited oscillation method as shown in Section 4.2.6.

Figure 4.33 illustrates an experimental setup that may be used to identify the approximate transfer function of a servo valve by the self-excited oscillation method shown schematically in the block diagram in Fig. 4.25. In the experimental setup, the on–off element comprises an electronic

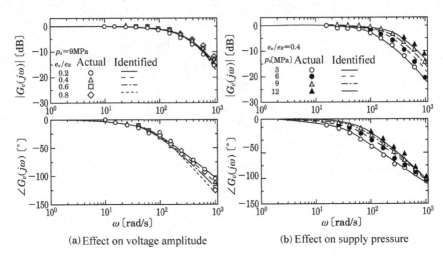

(a) Effect on voltage amplitude   (b) Effect on supply pressure

Fig. 4.34   Frequency responses of a servo valve.

comparator and electronic relays. Angular frequency ω_s and amplitude V_A measured in the self-excited oscillation yield the undamped natural angular frequency ω_n and damping coefficient ζ of the approximate transfer function that is obtained by the identification procedures outlined in Section 4.2.6. In Fig. 4.34, the actual servo valve frequency response is compared with that of the approximate transfer function derived by the self-excited oscillation method. This is based on the assumption that the input amplitude V_A in the frequency response corresponds to the value of e_a in the on–off element of the self-excited system in Fig. 4.34. Figure 4.34 demonstrates that the actual frequency characteristics almost agree with the frequency response of the identified transfer function in the range from $\omega \approx 0$ to the frequency at the phase angle $\phi \approx 80°$.

Figure 4.35 shows the approximate servo valve characteristics (i.e. undamped natural angular frequencies ω_{nv} and damping coefficients ζ_v) identified by the self-excited oscillation method for a range of supply pressures.

4.3.3 *Dynamic Performance of a Hydraulic System*

This section deals with the performance of a hydraulic actuator system with a servo valve. The actuator response considered here is derived for the case when the working fluid inertia and the hydraulic friction loss between

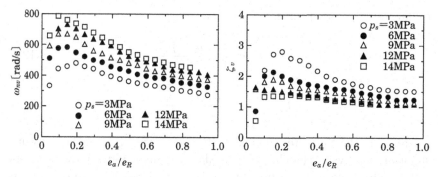

Fig. 4.35 Undamped natural angular frequencies and damping coefficients of a servo valve.

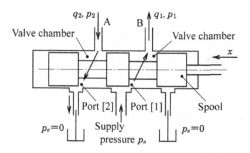

Fig. 4.36 Flows in a zero-lapped spool valve.

the servo valve and the actuator can be considered to be negligible, and moreover, the flow rate coefficient does not vary with the opening port area.

4.3.3.1 Flow Rate through a Spool Valve

(a) Flow rate equation of a zero lapped spool valve

When a zero lapped spool is located at the neutral position, the spool lands shut the valve ports, and then it is assumed that the flow leakage will not occur. When the spool is displaced from the central position by a distance x, as shown in Fig. 4.36, the opening area a of the spool port with width w is denoted as $a = wx$. Thus, the hydraulic fluid flows into the right-hand side valve chamber through port [1] at the rate q_1, and the fluid in the left valve chamber flows out through port [2] at the rate q_2. Applying Eq. (2.59) to flow rates q_1 and q_2 gives

$$q_1 = C_d wx \sqrt{\frac{2(p_s - p_1)}{\rho}}, \tag{4.123}$$

$$q_2 = C_d wx \sqrt{\frac{2(p_2 - p_e)}{\rho}}, \tag{4.124}$$

where p_s is the pressure supplied to the servo valve, p_1 is the pressure supplied to the valve chamber on one-side, p_2 is the pressure in the exhaust-end chamber, and p_e is the pressure in the tank, which is regarded as zero.

Flow rate q_1 is assumed to be identical to q_2 because the fluid leakages in the spool chambers are negligibly small. For pressure $p_1 = p_{10}$ and $p_2 = p_{20}$ in the equilibrium state,

$$q_1 = q_2 = q_0. \tag{4.125}$$

Equations (4.123), (4.124) and (4.125) give

$$p_{10} + p_{20} = p_s. \tag{4.126}$$

Let us denote differential pressure $p_{10} - p_{20}$ as $p_{\ell 0}$

$$p_{\ell 0} = p_{10} - p_{20}. \tag{4.127}$$

Equations (4.126) and (4.127) give

$$p_{10} = \frac{p_s + p_{\ell 0}}{2}, \quad p_{20} = \frac{p_s - p_{\ell 0}}{2}. \tag{4.128}$$

Substituting Eq. (4.128) into Eq. (4.123) gives the flow rate q_0 for $x = x_0$

$$q_0 = C_d wx_0 \sqrt{\frac{p_s - p_{\ell 0}}{\rho}}. \tag{4.129}$$

(b) *Flow rate equation of an under lapped spool valve*

It is a well-established fact that the dynamic performance of an under lapped spool valve is superior to a zero lapped spool valve in terms of the stability of the spool movement around the neutral position. Figure 4.37 illustrates an under lapped spool that has a clearance c in the neutral spool position where leakage flow rates $q_3 = q_4$ and $q_5 = q_6$ occur in the valve chambers even though the flow rate q_1 into the actuator and the flow rate q_2 out of the actuator are zero.

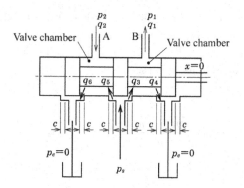

Fig. 4.37 Flows in an under lapped spool valve.

When the spool displacement x from the neutral position is less than the clearance c, the flow rate q_1 into the actuator and the flow rate q_2 out of the actuator in Fig. 4.37 are defined as follows:

$$\left.\begin{array}{c} q_1 = q_3 - q_4 \\ q_2 = q_6 - q_5 \end{array}\right\}. \tag{4.130}$$

Recalling Eq. (2.59), Eq. (4.130) becomes as follows. For $|x| \le c$

$$q_1 = C_d w(c + x)\sqrt{\frac{2(p_s - p_1)}{\rho}} - C_d w(c - x)\sqrt{\frac{2p_1}{\rho}}, \tag{4.131}$$

$$q_2 = C_d w(c + x)\sqrt{\frac{2p_2}{\rho}} - C_d w(c - x)\sqrt{\frac{2(p_s - p_2)}{\rho}}, \tag{4.132}$$

where spool displacement x is defined as having a plus sign in the left direction from the neutral position. The area of the opening spool port a is expressed as $a = w(x + c)$ for $|x| > c$, and then the leakage flow rates q_4 and q_5 become zero. Therefore, the flow rates q_1 and q_2 can be given as follows. For $|x| > c$

$$q_1 = C_d w(x + c)\sqrt{\frac{2(p_s - p_1)}{\rho}}, \tag{4.133}$$

$$q_2 = C_d w(x + c)\sqrt{\frac{2p_2}{\rho}}. \tag{4.134}$$

Equation (4.128) is also applicable to an under lapped spool valve, because Eq. (4.126) is derived by setting the first term in the right side of Eq. (4.131) equal to the first term in the right-hand side of Eq. (4.132), and hence the second term of Eq. (4.131) becomes equal to the second term in Eq. (4.132). By virtue of Eq. (4.128), Eqs. (4.131) and (4.133) yield the flow rate q_0 through the servo valve as follows.

For $|x| \leq c$:

$$q_0 = C_d w(c + x_0)\sqrt{\frac{p_s - p_{\ell 0}}{\rho}} - C_d w(c - x_0)\sqrt{\frac{p_s + p_{\ell 0}}{\rho}}. \tag{4.135}$$

For $|x| > c$:

$$q_0 = C_d w(x_0 + c)\sqrt{\frac{p_s - p_{\ell 0}}{\rho}}. \tag{4.136}$$

(c) *Flow rate equation of a servo valve*

In a steady state condition, the input current i_s from the servo amplifier is proportional to the displacement x_0 of servo valve spool. Thus, the flow rate equation corresponding to Eq. (4.129) may be written as follows:

$$q_s = C i_s \sqrt{p_s - p_{\ell 0}}, \tag{4.137}$$

where p_s is the pressure supplied to the servo valve, and $p_{\ell 0}$ is the differential pressure between the inlet and the outlet of the actuator. Servo valve capacity factor C is estimated by substituting the rated supply pressure $p_s = p_{sR}$ and the rated no-load flow rate $q = q_R$ (given in the manufacturer's specification) into Eq. (4.137). Recalling Eqs. (4.123), (4.124), (4.128), (4.137) yields

$$q_{10} = q_{20} = \sqrt{2} C i_s \sqrt{p_s - p_{10}} = \sqrt{2} C i_s \sqrt{p_{20}}. \tag{4.138}$$

Using the comprehensive coefficient C_0, Eq. (4.138) is rearranged as follows:

$$\left. \begin{array}{l} q_1 = C_0 i \sqrt{p_s - p_1} \\ q_2 = C_0 i \sqrt{p_2} \\ C_0 = \sqrt{2} C \end{array} \right\}. \tag{4.139}$$

Figure 4.38 shows the block diagram of the signal transmission process taking into account the servo valve response lag. The input voltage $\Delta V(s)$ is transformed by a servo amplifier into the electric current $\Delta I(s) = K_a \Delta V(s)$ and it induces the spool displacement as in a standard second order lag

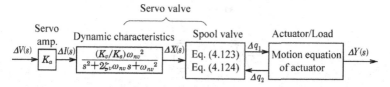

Fig. 4.38 Signal transmission process in a hydraulic servo system.

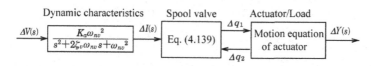

Fig. 4.39 Arranged signal transmission process in a hydraulic servo system.

system with the gain constant K_v/K_s. Thereby, the servo valve supplies the flow rate Δq_1 into an actuator, in accordance with Eq. (4.123), and instantaneously displaces the flow rate Δq_2 out of the actuator according as Eq. (4.124), so that the actuator moves with the load.

Taking into account the relationship between the output $\Delta Y(s)$ and input $\Delta V(s)$, Fig. 4.39 is equivalent to Fig. 4.38 because the servo valve's response lag performance, neglected in Eq. (4.139), is considered in Fig. 4.39 by adding a fictitious servo amplifier response time lag modifying the gain constants.

4.3.3.2 *Motion Equation of an Actuator*

(a) *Hydraulic cylinder*

Figure 4.40 shows an open loop hydraulic servo cylinder system. When the differential pressure $p_1 - p_2$ between the two cylinder chambers is generated, the hydraulic force $A(p_1 - p_2)$ moves the piston with area A against a load consisting of mass m, the viscous force with viscous coefficient B and Coulomb friction force f. The equation of motion of the piston is written as

$$\eta_c A(p_1 - p_2) = m\frac{d^2y}{dt^2} + B\frac{dy}{dt} + f, \tag{4.140}$$

where η_c denotes the thrust efficiency of the cylinder. As already explained in Section 2.3.2, the flow rate q_1 into the cylinder and the flow rate q_2 out

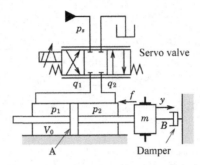

Fig. 4.40   Hydraulic cylinder system.

of the cylinder may be described as follows:

$$\left. \begin{aligned} q_1 &= A\frac{dy}{dt} + \frac{(V_0 + Ay)}{\kappa}\frac{dp_1}{dt} \\[2mm] q_2 &= A\frac{dy}{dt} - \frac{(V_0 - Ay)}{\kappa}\frac{dp_2}{dt} \end{aligned} \right\}. \tag{4.141}$$

For $V_0 \gg Ay$, Eq. (4.141) may be approximated as follows:

$$\left. \begin{aligned} q_1 &= A\frac{dy}{dt} + \frac{V_0}{\kappa}\frac{dp_1}{dt} \\[2mm] q_2 &= A\frac{dy}{dt} - \frac{V_0}{\kappa}\frac{dp_2}{dt} \end{aligned} \right\}, \tag{4.142}$$

where κ and V_0 denote the bulk modulus of the hydraulic fluid and one side cylinder chamber volume V_0 at the central piston position, respectively. When the values of the coefficients in these equations are provided, the simultaneous solution of Eqs. (4.139), (4.140) and (4.142) on variables y, p_1, p_2, q_1 and q_2 gives the piston behavior for the specified input electric current.

(b) *Hydraulic motor*

Figure 4.41 illustrates a hydraulic servo motor system. When the differential pressure $p_1 - p_2$ between the inlet and the outlet of the motor is induced by the hydraulic source system, the motor shaft torque T_{am} will rotate the load with a moment of inertia of J at an angular speed $\omega = d\theta/dt$, and then the fluid will flow at a flow rate of q through the servo valve. The power of the motor shaft is expressed as

$$T_{am}\omega = \eta_m q(p_1 - p_2), \tag{4.143}$$

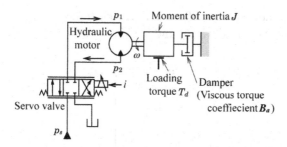

Fig. 4.41 Hydraulic motor system.

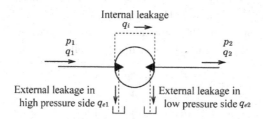

Fig. 4.42 Leakage flows in a hydraulic motor.

where η_{mT} and η_{mv} are the torque efficiency and the volumetric efficiency respectively, and $\eta_m = \eta_{mT}\eta_{mv}$ is the overall efficiency of the motor. Denoting a theoretical motor displacement per unit radian by D_m, the flow rate q into the motor is given as follows:

$$q = \frac{D_m\omega}{\eta_{mv}}. \tag{4.144}$$

Canceling angular velocity ω in Eqs. (4.143) and (4.144) gives

$$T_{am} = \eta_{mT}D_m(p_1 - p_2). \tag{4.145}$$

When the motor shaft rotates at angular velocity $\omega = d\theta/dt$ against the load, the equation of motion of the motor shaft is given as

$$\eta_{mT}D_m(p_1 - p_2) = J\frac{d^2\theta}{dt^2} + B_a\frac{d\theta}{dt} + T_d, \tag{4.146}$$

where J denotes the moment of inertia of the motor shaft with the load, B_a is a viscous torque coefficient and T_d is a constant load torque.

Figure 4.42 illustrates the inner leakage flow rate q_i, outer leakage flow rate q_{e1} at the motor inlet and outer leakage flow rate q_{e2} at the motor outlet. These leakage flow rates are proportional to the differences in pressures upstream and downstream of the leakage clearance, as shown in Hagen–Poiseuille equation in Eq. (2.75) and in Couette flow equation in Eq. (2.103)

$$\left.\begin{array}{l} q_i = \dfrac{1}{R_i}p_\ell \\[2ex] q_{e1} = \dfrac{1}{R_e}p_1 \\[2ex] q_{e2} = \dfrac{1}{R_e}p_2 \end{array}\right\}, \tag{4.147}$$

where R_i and R_e are the leakage flow resistances. Taking account of the leakage flows in the flow into the hydraulic motor at the flow rate q_1 and the flow rate q_2 out of it, the applicable flow rate equations Eqs. (2.69) and (2.70) for flow rates q_1 and q_2 are defined as follows:

$$q_1 = D_m\frac{d\theta}{dt} + \frac{V_0}{\kappa}\frac{dp_1}{dt} + \frac{1}{R_i}p_\ell + \frac{1}{R_e}p_1, \tag{4.148}$$

$$q_2 = D_m\frac{d\theta}{dt} - \frac{V_0}{\kappa}\frac{dp_2}{dt} + \frac{1}{R_i}p_\ell - \frac{1}{R_e}p_2, \tag{4.149}$$

where κ, D_m and V_0 denote the bulk modulus of the hydraulic fluid, motor displacement per radian and the one side volume of the pipe/motor chamber, respectively. Simultaneous equations, Eqs. (4.139), (4.146) to Eq. (4.149) being solved by computer analysis, the cylinder response behaviors can be simulated for the specified input.

4.3.4 *Approximate Transfer Function of an Open Loop Hydraulic Servo System*

4.3.4.1 *Hydraulic Servo Cylinder System*

(a) *Servo system with a load consisting of a mass and viscous resistance force*

Though Eq. (4.139) is a nonlinear function, it may be linearized in the same manner as shown in Eqs. (4.111) and (4.112) as long as the flow rate changes only slightly from the steady state value. When pressures $p_1 = p_{10}$ and $p_2 = p_{20}$ change respectively to $p_1 = p_{10} + \Delta p_1$ and $p_2 = p_{20} + \Delta p_2$ for the small input change Δi, the flow rates $q_1 = q_{10}$ and $q_2 = q_{20}$ will change to $q_1 = q_{10} + \Delta q_1$ and $q_2 = q_{20} + \Delta q_2$ at the same time. Expanding Eq. (4.139) in the form of a Taylor's series and neglecting the terms with higher exponents than two, the equations of flow rate changes Δq_1 and Δq_2

are given as follows:

$$\Delta q_1 = k_{1i}\Delta i - k_{1p}\Delta p_1, \qquad (4.150)$$

$$\Delta q_2 = k_{2i}\Delta i + k_{2p}\Delta p_2, \qquad (4.151)$$

where

$$\left. \begin{array}{l} k_{1i} = \dfrac{\partial q_1}{\partial i}\bigg]_{p_{10}} = C_0\sqrt{p_s - p_{10}} \\[3mm] k_{1p} = -\dfrac{\partial q_1}{\partial p_1}\bigg]_{p_{10},\,i_0} = \dfrac{C_0 i_0}{2\sqrt{p_s - p_{10}}} \end{array} \right\}, \qquad (4.152)$$

$$\left. \begin{array}{l} k_{2i} = \dfrac{\partial q_2}{\partial i}\bigg]_{p_{20}} = C_0\sqrt{p_{20}} \\[3mm] k_{2p} = \dfrac{\partial q_2}{\partial p_2}\bigg]_{p_{20},\,i_0} = \dfrac{C_0 i_0}{2\sqrt{p_{20}}} \end{array} \right\}. \qquad (4.153)$$

Introducing $k_{1i} = k_{2i}$ and $k_{1p} = k_{2p}$ obtained by substituting Eq. (4.126) into Eq. (4.152), the flow rate change Δq_m is given as the following equation:

$$\Delta q_m = \frac{\Delta q_1 + \Delta q_2}{2} = k_1\Delta i - k_2\Delta p_\ell. \qquad (4.154)$$

$$\left. \begin{array}{l} k_1 = C_0\sqrt{p_s - p_{10}} = C\sqrt{p_s - p_{\ell 0}} \\[3mm] k_2 = \dfrac{C_0 i_0}{4\sqrt{p_s - p_{10}}} = \dfrac{C i_0}{2\sqrt{p_s - p_{\ell 0}}} \\[3mm] C_0 = \sqrt{2}C \end{array} \right\}. \qquad (4.155)$$

Recalling Eq. (4.142) applicable to flows into and out of the cylinder, gives the flow rate change $\Delta q_m = (\Delta q_1 + \Delta q_2)/2$ as follows:

$$\Delta q_m = A\Delta u + \frac{V_0}{2\kappa}\frac{d\Delta p_\ell}{dt}, \qquad (4.156)$$

where Δp_ℓ denotes a small change of the differential pressure $p_\ell = p_1 - p_2$ between the two cylinder chambers, and $\Delta u = d\Delta y/dt$ denotes a small change in the piston velocity for input current change Δi. The equation of motion for $f = 0$ in Eq. (4.140) is given as

$$\eta_c A\Delta p_\ell = m\frac{d\Delta u}{dt} + B\Delta u. \qquad (4.157)$$

When all of the initial conditions $\Delta q_m(0)$, $\Delta p_l(0)$, $\Delta u(0)$, $d\Delta p_l(0)/dt$ and $d\Delta u(0)/dt$ are zero, the Laplace transforms of Eqs. (4.154), (4.156) and

(4.157) are given as:

$$
\begin{bmatrix}
1 & k_2 & 0 \\
1 & -(V_0/2\kappa)s & -A \\
0 & \eta_c A & -(ms+B)
\end{bmatrix}
\begin{bmatrix}
\Delta Q_m(s) \\
\Delta P_\ell(s) \\
\Delta U(s)
\end{bmatrix}
=
\begin{bmatrix}
k_1 \\
0 \\
0
\end{bmatrix}
\Delta I(s), \qquad (4.158)
$$

where $\Delta Q_m(s), \Delta P_\ell(s), \Delta U(s)$ and $\Delta I(s)$ denote Laplace transforms of time-variable functions $\Delta q_m(t)$, $\Delta p_\ell(t)$, $\Delta u(t)$ and $\Delta i(t)$, respectively. Canceling $\Delta Q_m(s)$ and $\Delta P_\ell(s)$ in Eq. (4.158) gives:

$$
G_u(s) = \frac{\Delta U(s)}{\Delta I(s)} = \frac{\eta_c A k_1}{\left(k_2 + \frac{V_0}{2\kappa}s\right)(ms+B) + \eta_c A^2} = \frac{K_u \omega_n^2}{s^2 + 2\zeta\omega_n s + \omega_n^2},
$$
$$(4.159)$$

where

$$
\left.
\begin{aligned}
K_u &= \frac{\eta_c k_1 A}{B k_2 + \eta_c A^2} \\[2mm]
\omega_n &= \sqrt{\frac{2\kappa(k_2 B + \eta_c A^2)}{V_0 m}} \\[2mm]
\zeta &= \sqrt{\frac{V_0 m}{2\kappa(k_2 B + \eta_c A^2)}} \left(\frac{B}{2m} + \frac{\kappa k_2}{V_0}\right)
\end{aligned}
\right\}. \qquad (4.160)
$$

When the value of V_0/κ is so small that the term $(V_0/2\kappa)(d\Delta p/dt)$ in Eq. (4.156) can be neglected, Eqs. (4.159) and (4.160) are rearranged as follows:

$$
\left.
\begin{aligned}
G_u(s) &= \frac{\Delta U(s)}{\Delta I(s)} = \frac{K_u}{Ts+1} \\[2mm]
K_u &= \frac{\eta_c k_1 A}{B k_2 + \eta_c A^2} \\[2mm]
T &= \frac{m k_2}{B k_2 + \eta_c A^2}
\end{aligned}
\right\}. \qquad (4.161)
$$

However, this term is not negligible except for very small load conditions. Therefore, Eq. (4.161) may turn out to be inapplicable in most cases.

Equations (4.155) and (4.160) show that the damping ratio ζ in $G_u(s)$ is the smallest at the neutral spool position for the zero lapped spool valve because the coefficient k_2 is zero whilst the input current $i_0 = 0$. Therefore, the stability performance of the servo system tends to deteriorate in the neighborhood of the neutral spool position.

Let us now consider the servo system with an under lapped spool. Replacing the spool displacement x in Eq. (4.131) and (4.132) by the equivalent electric current i, the flow rates q_1 and q_2 through the servo valve with an under lapped length c are given as follows:

For $i \leq i_c$:

$$q_1 = C_0(i_c + i)\sqrt{p_s - p_1} - C_0(i_c - i)\sqrt{p_1}, \qquad (4.162)$$

$$q_2 = C_0(i_c + i)\sqrt{p_2} - C_0(i_c - i)\sqrt{p_s - p_2}, \qquad (4.163)$$

where i_c denotes the input electrical current corresponding to the under lapped length c, and C_0 is the coefficient expressing the servo valve capacity. Applying Eqs. (4.162) and (4.163) to the flow rate change Δq_m, we get

$$\Delta q_m = \frac{\Delta q_1 + \Delta q_2}{2}$$

$$= \frac{1}{2}\left(\left[\frac{\partial q_1}{\partial i}\right]_{p_{10}} + \left[\frac{\partial q_2}{\partial i}\right]_{p_{20}}\right)\Delta i$$

$$+ \frac{1}{2}\left(\left[\frac{\partial q_1}{\partial p_1}\right]_{p_{10},i_0}\Delta p_1 + \left[\frac{\partial q_2}{\partial p_2}\right]_{p_{20},i_0}\Delta p_2\right)$$

$$= \frac{1}{2}C_0(\sqrt{p_s - p_{10}} + \sqrt{p_{10}} + \sqrt{p_{20}} + \sqrt{p_s - p_{20}})\Delta i$$

$$- \frac{C_0}{4}\left(\frac{i_c + i_0}{\sqrt{p_s - p_{10}}} + \frac{i_c - i_0}{\sqrt{p_{10}}}\right)\Delta p_1$$

$$+ \frac{C_0}{4}\left(\frac{i_c + i_0}{\sqrt{p_{20}}} + \frac{i_c - i_0}{\sqrt{p_s - p_{20}}}\right)\Delta p_2. \qquad (4.164)$$

Substituting Eq. (4.128) into Eq. (4.164) gives

$$\Delta q_m = k_1\Delta i - k_2\Delta p_\ell, \qquad (4.165)$$

$$\left.\begin{array}{l} k_1 = C\left(\sqrt{p_s - p_{\ell 0}} + \sqrt{p_s + p_{\ell 0}}\right) \\[2mm] k_2 = \dfrac{C}{2}\left(\dfrac{i_c + i_0}{\sqrt{p_s - p_{\ell 0}}} + \dfrac{i_c - i_0}{\sqrt{p_s + p_{\ell 0}}}\right) \end{array}\right\}, \qquad (4.166)$$

$$C = \frac{C_0}{\sqrt{2}}. \qquad (4.167)$$

As shown in Eqs. (4.154) and (4.165), the flow rate changes Δq_m are expressed by the same equation for both valves with the zero lapped spool and the under lapped spool, though the relevant coefficients are expressed differently. Therefore, the servo system transfer function $G_u(s)$ given by Eq. (4.159) is applicable to a servo system with an under lapped spool, provided that relevant values of the coefficients in Δq_m are selected depending on the spool type.

Equations (4.166) shows that the under lapped spool improves the stability performance of the servo system in the neighborhood of the neutral position, because the coefficient k_2 is not zero at $i_0 = 0$ for that type of spool. By virtue of Eqs. (4.159) and (4.121) and recalling Fig. 4.39, the open-loop transfer function $\Delta U(s)/\Delta V(s)$ is written as

$$\frac{\Delta U(s)}{\Delta V(s)} = \frac{K_a \omega_{nv}^2}{s^2 + 2\zeta_v \omega_{nv} s + \omega_{nv}^2} G_u(s), \qquad (4.168)$$

where $\Delta U(s)$ denotes the Laplace transform of the piston output speed change $\Delta u = d\Delta y/dt$, and $\Delta V(s)$ denotes the transform of the input voltage change Δv.

(b) *Hydraulic system with a load consisting of a mass, viscous resistance and restoring force*

Figure 4.43 illustrates a cylinder whose piston rod end is connected to an aircraft control plate. When the piston moves from the reference position by the distance Δy, the arm with length L makes the control plate incline at angle $\Delta \delta$ around the rotation center O. The equation of motion of the piston can be written as

$$\eta_c A \Delta p_\ell - \Delta f = m \frac{d^2 \Delta y}{dt^2} + B \frac{d\Delta y}{dt}, \qquad (4.169)$$

where Δp_ℓ is the differential pressure change between the cylinder chambers, Δf is the change of the force delivered by the control plate, B is the

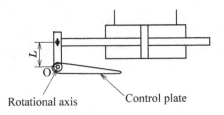

Fig. 4.43   Cylinder load.

coefficient of the viscous resistance acting on the piston and η_c is the thrust efficiency of the cylinder. Thus, the motion equation of the control plate is given as follows:

$$\Delta f L = J \frac{d^2 \Delta \delta}{dt^2} + B_\delta \frac{d\Delta \delta}{dt} + k_\delta \Delta \delta, \tag{4.170}$$

$$\Delta \delta = \frac{\Delta y}{L}, \tag{4.171}$$

where J is the moment of inertia around the center O, B_δ is the coefficient of the viscous resistance torque and k_δ is the coefficient of the restoring torque produced by air resistance during the flight. Assuming all the initial conditions to be zero, the Laplace transforms of Eqs. (4.154), (4.156) (4.169), (4.170) and (4.171) are given as

$$\begin{bmatrix} 1 & k_2 & 0 \\ 1 & -(V_0/2\kappa)s & -ALs \\ 0 & \eta_c LA & -(mL^2 + J)s^2 - (BL^2 + B_\delta)s - k_\delta \end{bmatrix} \begin{bmatrix} \Delta Q_m(s) \\ \Delta P_\ell(s) \\ \Delta \delta(s) \end{bmatrix}$$

$$= \begin{bmatrix} k_1 \\ 0 \\ 0 \end{bmatrix} \Delta I(s). \tag{4.172}$$

Canceling $\Delta Q_m(s)$ and $\Delta P_\ell(s)$ in Eq. (4.172) gives

$$\left. \begin{aligned} G_\delta(s) = \frac{\Delta \delta(s)}{\Delta I(s)} &= \frac{b}{s^3 + a_1 s^2 + a_2 s + a_3} \\ a_1 &= \frac{2k_2 \kappa}{V_0} + \frac{BL^2 + B_\delta}{mL^2 + J} \\ a_2 &= \frac{2\eta_c \kappa A^2 L^2 + 2\kappa k_2(BL^2 + B_\delta) + V_0 k_\delta}{V_0(mL^2 + J)} \\ a_3 &= \frac{2\kappa k_2 k_\delta}{V_0(mL^2 + J)} \\ b &= \frac{2\eta_c \kappa A L k_1}{V_0(mL^2 + J)} \end{aligned} \right\}. \tag{4.173}$$

Considering $\Delta I(s)/\Delta V(s) = K_a\omega_{nv}^2/(s^2 + 2\zeta_v\omega_{nv}s + \omega_{nv}^2)$ with reference to Fig. 4.39, the open loop transfer function $\Delta\delta(s)/\Delta V(s)$ is written as

$$\frac{\Delta\delta(s)}{\Delta V(s)} = \frac{K_a\omega_{nv}^2}{s^2 + 2\zeta_v\omega_{nv}s + \omega_{nv}^2}G_\delta(s). \tag{4.174}$$

(c) *Open loop hydraulic servo motor system*

Applying Eqs. (4.148) and (4.149) with reference to the flow rate $q_m = (q_1 + q_2)/2$ gives

$$\left.\begin{array}{c} q_m = D_m\dfrac{d\theta}{dt} + \dfrac{V_0}{2\kappa}\dfrac{dp_\ell}{dt} + \dfrac{1}{R}p_\ell \\[2ex] \dfrac{1}{R} = \dfrac{1}{R_i} + \dfrac{1}{2R_e} \end{array}\right\}. \tag{4.175}$$

Equations (4.146) and (4.175) may be expressed in terms of infinitesimally changing variables

$$\eta_{mT}D_m\Delta p_\ell = J\frac{d\Delta\omega}{dt} + B_a\Delta\omega, \tag{4.176}$$

$$\Delta q_m = D_m\Delta\omega + \frac{V_0}{2\kappa}\frac{d\Delta p_\ell}{dt} + \frac{1}{R}\Delta p_\ell, \tag{4.177}$$

where the extraneous torque T_d in Eq. (4.146) is assumed to be zero.

Assuming all of the initial conditions $\Delta q_m(0), \Delta p_\ell(0), \Delta\omega(0)$, $d\Delta p_\ell(0)/dt$ and $d\Delta\omega(0)/dt$ to be zero, the Laplace transforms of Eqs. (4.154), (4.176) and (4.177) are expressed as

$$\begin{bmatrix} 1 & k_2 & 0 \\ 1 & -(V_0/2\kappa)s - (1/R) & -D_m \\ 0 & \eta_{mT}D_m & -Js - B_a \end{bmatrix}\begin{bmatrix} \Delta Q_m(s) \\ \Delta P_\ell(s) \\ \Delta\omega(s) \end{bmatrix} = \begin{bmatrix} k_1 \\ 0 \\ 0 \end{bmatrix}\Delta I(s). \tag{4.178}$$

Canceling $\Delta Q_m(s)$ and $\Delta P_\ell(s)$ in Eq. (4.178) gives

$$\begin{aligned} G_\omega(s) &= \frac{\Delta\omega(s)}{\Delta I(s)} = \frac{\eta_{mT}D_mk_1}{\left(k_2 + \frac{1}{R} + \frac{V_0}{2\kappa}s\right)(Js + B_a) + \eta_{mT}D_m^2} \\[2ex] &= \frac{K_\omega\omega_n^2}{s^2 + 2\zeta\omega_ns + \omega_n^2}, \end{aligned} \tag{4.179}$$

where the gain constant K_ω, undamped natural frequency ω_n and damping coefficient ζ are as follows:

$$K_\omega = \frac{\eta_{mT} k_1 D_m}{B_a \left(k_2 + \frac{1}{R}\right) + \eta_{mT} D_m^2}, \tag{4.180}$$

$$\left.\begin{aligned}
\omega_n &= \sqrt{\frac{2\kappa \left[B_a \left(k_2 + \frac{1}{R}\right) + \eta_{mT} D_m^2\right]}{V_0 J}} \\
\zeta &= \left[\frac{B_a}{2J} + \frac{\kappa}{V_0}\left(\frac{1}{R} + k_2\right)\right]\sqrt{\frac{V_0 J}{2\kappa\left[B_a\left(k_2 + \frac{1}{R}\right) + \eta_{mT} D_m^2\right]}}
\end{aligned}\right\}. \tag{4.181}$$

4.3.5 *Identification of an Open Loop Transfer Function*

In the synthesis of the hydraulic servo system, the open loop transfer function is of primary importance. Figure 4.44(a) illustrates a block diagram of the hydraulic servo system comprising a servo amplifier, a servo valve, a hydraulic motor/load, a sensor and a feedback signal. The response speeds

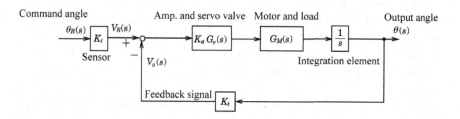

(a) Hydraulic angular control system

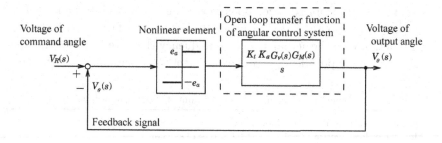

(b) Self-excited oscillation system

Fig. 4.44   Hydraulic angular control system and a self-excited oscillation set up.

of the sensor and the servo amplifier are both so high that each transfer function may be expressed in terms of gain constants K_a and K_t, respectively. Even though the transfer functions $G_v(s)$ of the servo valve and $G_M(s)$ of the motor/load system can be approximated well by the standard second order lag systems as shown in Eq. (4.121) and (4.179), the quantitative estimation of factors in the transfer functions is a formidable task. However, an open loop transfer function $K_t K_a G_v(s) G_M(s)/s$ as shown in Fig. 4.44(a) can often be approximated by a standard second-order lag system with an integral element, because the undamped natural frequency ω_{nv} of the servo valve is ten times or more than the undamped natural frequency ω_n of the motor/load system on the whole. Thus, the open loop transfer function $K_t K_a G_v(s) G_M(s)$ can be obtained as the approximate second order lag transfer function using the self-excited oscillation method outlined in Section 4.2.6.

Figure 4.44(b) shows the block diagram for an identification setup appending the nonlinear element to provide the bang-bang control input $\pm e_a$ to the hydraulic angular control system. The system enclosed by the dashed line is the objective servo valve/motor/load system which may be used to identify the approximated transfer function. The identification setup generates a self-excited oscillation with small amplitude, according to the bang-bang control error input e_a around the commanded angular position. The dimensional specifications of the set up are shown in Table 4.9.

Figure 4.45 shows the oscillation wave in a self-excited oscillation system. The waves are getting more and more distorted as the bang-bang control input e_a becomes large not appeared in this figure. The analysis of the frequency and amplitude of the oscillation in accordance with the identification procedure, shown in Section 4.2.6, yields the dynamic parameters ζ and ω_n in the approximated open loop transfer function to be identified.

In order to compare the frequency characteristics of the identified transfer function with the actual open loop frequency characteristics for the sinusoidal input amplitude $V_a = e_a$, experimental measurements of the open

Table 4.9   Dimensions of a hydraulic motor/load system.

One side volume of motor chamber	$V_0 = 16\,\text{cm}^3$
Moment of load inertia	$J = 0.0814\,\text{kg} \cdot \text{cm}^2$
Motor displacement	$D_m = 1.7\,\text{cm}^3/\text{rad}$
Capacity factor of servo valve	$C = 4.45 \times 10^{-9}\,\text{m}^3/(\text{mA} \cdot \text{s} \cdot \text{Pa}^{1/2})$
Viscous damping coefficient of load	$B_a = 70\,\text{kg}\cdot\text{cm}^2/(\text{s}\cdot\text{rad})$ to $100\,\text{kg}\cdot\text{cm}^2/(\text{s}\cdot\text{rad})$

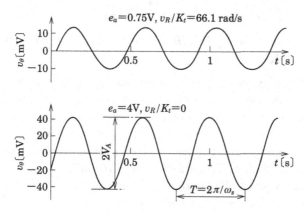

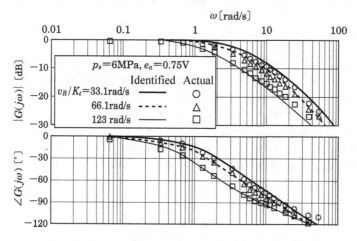

Fig. 4.45 Self excited oscillation waves.

Fig. 4.46 Frequency characteristics of a hydraulic motor/load system.

loop frequency responses are conducted. Though the frequency response waves are distorted as the sinusoidal input amplitude $V_a = e_a$ increases, the experimental frequency characteristics may be obtained by adopting the mean frequency characteristics between the peak and the trough of the wave. Figure 4.46 shows the comparisons of the actual frequency characteristics of the open loop system with those of the identified transfer functions. Both frequency characteristic curves agree well in the phase angle range from 0 to about $-90°$.

4.4 Design of a Hydraulic Control System

4.4.1 *Capacity of a Hydraulic Component Selected to Meet Power Requirements*

When designing hydraulic servo mechanisms, it is necessary to determine operational parameters such as the supply pressure, the flow capacity of the servo valve and the actuator dimension, so as to satisfy the power requirements for the given working application. It is usual to draw a load chart that designates the required power in terms of energy saving. This chart gives the relationship between load force and velocity. The driving characteristic curve of the hydraulic system should effectively enclose the locus of the load chart with a small margin.

In this section, the load chart corresponding to the demanded response speed is explained, with the main focus being on how the load chart should be fitted in the driving characteristic curve of the hydraulic system.

4.4.1.1 *Load Chart*

Figure 4.47 illustrates a hydraulic servo system with a load that consists of a mass, a damping force and a spring force. Let us consider the load chart of the system responding at the equivalent time constant T_e to the step input to produce the piston displacement $y = a$.

Since the inverse of the equivalent time constant T_e corresponds to the bandwidth ω_b, as shown in Table 4.7, it is required that the system should have the ability to drive the load at the sinusoidal motion $y = a \sin \omega_b t$.

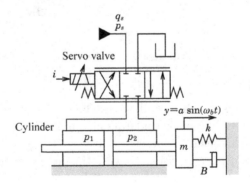

Fig. 4.47   Hydraulic cylinder/load system.

Therefore, the load conditions can be defined as

$$\omega = \omega_b = \frac{1}{T_e},\qquad(4.182)$$

$$y = a\sin\omega_b t,\qquad(4.183)$$

$$v = \frac{dy}{dt} = a\omega_b\cos\omega_b t.\qquad(4.184)$$

Thus, the load force for the sinusoidal motion is given as

$$f = m\frac{d^2y}{dt^2} + B\frac{dy}{dt} + ky = -a(m\omega_b^2 - k)\sin\omega_b t + Ba\omega_b\cos\omega_b t,\qquad(4.185)$$

where m, B and k denote the load mass, the viscous force coefficient and the spring constant, respectively and it is supposed that the inertia force $am\omega_b^2$ is significantly larger than the spring force ak. The dependence between the load force f and load velocity v is derived from Eq. (4.182) to Eq. (4.185)

$$v^2 + \alpha f^2 - 2\beta v f = \gamma,\qquad(4.186)$$

$$\left.\begin{array}{l}\alpha = \dfrac{\omega_b^2}{B^2\omega_b^2 + (m\omega_b^2 - k)^2}\\[2ex]\beta = \dfrac{B\omega_b^2}{B^2\omega_b^2 + (m\omega_b^2 - k)^2}\\[2ex]\gamma = \dfrac{a^2\omega_b^2(m\omega_b^2 - k)^2}{B^2\omega_b^2 + (m\omega_b^2 - k)^2}\end{array}\right\}.\qquad(4.187)$$

Equation (4.186) yields an elliptical load chart as shown in Fig. 4.48. The semi-major axis a^*, semi-minor axis b^* and the incidence angle θ are

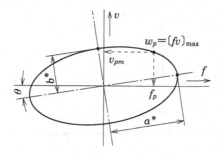

Fig. 4.48   Load chart.

denoted [See Note] as in Eq. (4.188) assuming that the scale of the velocity is n-times the force scale.

$$\left.\begin{array}{l} \theta = \dfrac{1}{2}\tan^{-1}\dfrac{-2n\beta}{\alpha-n^2} = \dfrac{1}{2}\tan^{-1}\dfrac{-2nB\omega_b^2}{-n^2(m\omega_b^2-k)^2+\omega_b^2(1-n^2B^2)} \\[4mm] a^* = \sqrt{\dfrac{\gamma}{n^2\sin^2\theta+\alpha\cos^2\theta-2n\beta\sin\theta\cos\theta}} \\[4mm] b^* = n\sqrt{\dfrac{\gamma}{n^2\cos^2\theta+\alpha\sin^2\theta+2n\beta\sin\theta\cos\theta}} \end{array}\right\}.$$

$$(4.188)$$

==

[Note] Concerning the elliptical locus in Fig. 4.48.

Recalling a rectangular coordinate X-Y with an incident angle θ to a rectangular coordinate x-y as shown in Fig. 4.49, the relationships between the two coordinates are given as

$$\left.\begin{array}{l} x = X\cos\theta + (Y/n)\sin\theta \\[2mm] y = -nX\sin\theta + Y\cos\theta \end{array}\right\}.$$

$$(1)$$

The equation of an ellipse with a semi-major axis a^* and semi-minor axis b^* in the coordinate x-y is given as follows:

$$\frac{x^2}{a^{*2}} + \frac{y^2}{b^{*2}} = 1.$$

$$(2)$$

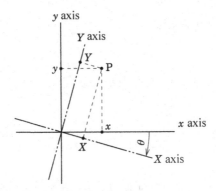

Fig. 4.49   Relationships between the rectangular coordinates x-y and X-Y.

Substituting Eq. (1) into Eq. (2) gives

$$Y^2 + aX^2 - 2\beta XY = \gamma, \tag{3}$$

where the coefficients α, β and γ are as follows:

$$\left.\begin{array}{l} \alpha = \dfrac{n^2 b^{*2} \cos^2\theta + n^4 a^{*2} \sin^2\theta}{b^{*2} \sin^2\theta + a^{*2} n^2 \cos^2\theta} \\[4mm] \beta = -\dfrac{nb^{*2}\cos\theta\sin\theta - n^3 a^{*2}\cos\theta\sin\theta}{b^{*2}\sin^2\theta + a^{*2} n^2 \cos^2\theta} \\[4mm] \gamma = \dfrac{n^2 a^{*2} b^{*2}}{b^{*2}\sin^2\theta + a^{*2} n^2 \cos^2\theta} \end{array}\right\}. \tag{4}$$

Equation (3) corresponds to Eq. (4.186). Equation (4) leads to the following equations:

$$\theta = \frac{1}{2}\tan^{-1}\frac{-2n\beta}{\alpha - n^2}, \tag{5}$$

$$a^* = \sqrt{\frac{\gamma}{n^2 \sin^2\theta + \alpha\cos^2\theta - 2n\beta\cos\theta\sin\theta}}, \tag{6}$$

$$b^* = n\sqrt{\frac{\gamma}{n^2 \cos^2\theta + \alpha\sin^2\theta + 2n\beta\cos\theta\sin\theta}}. \tag{7}$$

Therefore, the ellipse given by Eq. (2) in the rectangular coordinate x–y corresponds to Eq. (3) in the X–Y coordinate.

==

Recalling Eqs. (4.184) and (4.185), load power $w = fv$ is obtained as follows:

$$w = -\frac{1}{2}a^2\omega_b(m\omega_b^2 - k)\sin 2\omega_b t + a^2\omega_b^2 B\cos^2\omega_b t. \tag{4.189}$$

Replacing the differential in Eq. (4.189) by zero i.e. $dw/dt = 0$, we get

$$\tan(2\omega_b t_p \pm N\pi) = -\frac{m\omega_b^2 - k}{B\omega_b}, \tag{4.190}$$

where t_p denotes the time when the load power reaches its local maximum (or minimum) in the frequency response waves, and N is an integer. Equation (4.190) gives time t_p at the maximum (or minimum) load power

$$t_p = \frac{N\pi - \phi}{2\omega_b},\tag{4.191}$$

$$\phi = \tan^{-1}\frac{m\omega_b^2 - k}{B\omega_b}.\tag{4.192}$$

Substituting Eq. (4.191) into Eqs. (4.184) and (4.185), the load velocity v_p and load force f_p under the maximum power conditions are obtained accordingly:

$$v_p = a\omega_b \cos\frac{\phi}{2},\tag{4.193}$$

$$f_p = a(m\omega_b^2 - k)\sin\frac{\phi}{2} + a\omega_b B \cos\frac{\phi}{2}.\tag{4.194}$$

Replacing the differential in Eq. (4.185) by zero, time $t = t_m$ at the maximum load force is obtained by

$$t_m = \frac{N\pi - \phi}{\omega_b}.\tag{4.195}$$

Substituting Eq. (4.195) into Eq. (4.185) gives the maximum load force f_m:

$$f_m = a(m\omega_b^2 - k)\sin\phi + Ba\omega_b \cos\phi = a\sqrt{(m\omega_b^2 - k)^2 + (B\omega_b)^2}.\tag{4.196}$$

While Eq. (4.184) gives the maximum load speed v_m as follows:

$$v_m = a\omega_b.\tag{4.197}$$

4.4.1.2 *Driving Characteristic Curve of the Hydraulic System*

Recalling Eq. (4.137), a flow rate q_s through the servo valve is given as

$$q_s = Ci\sqrt{p_s - p_\ell},\tag{4.198}$$

where $p_\ell = p_1 - p_2$ denotes the differential pressure between the cylinder chambers.

Denoting the piston area, the volumetric cylinder efficiency and the thrust efficiency as A, η_v and η_c, respectively, the piston speed V and the piston thrust F are described as follows:

$$V = \frac{\eta_v q_s}{A}, \tag{4.199}$$

$$F = \eta_c A p_\ell. \tag{4.200}$$

Equations (4.198), (4.199) and (4.200) yield the driving characteristic equation for the rated electrical current $i = I_R$

$$V = \frac{\eta_v C I_R}{A\sqrt{A\eta_c}} \sqrt{A\eta_c p_s - F}. \tag{4.201}$$

Substituting $i = I_R$ and $p_\ell = 0$ into Eq. (4.198), the rated maximum flow rate q_0 for the no-load condition is given as

$$q_0 = C I_R \sqrt{p_s}, \tag{4.202}$$

where $C I_R$ is the capacity factor of the servo valve. Equation (4.201) gives a parabolic driving characteristic curve in the rectangular coordinate system of abscissa F and ordinate V. Figure 4.50 illustrates the outline of the driving characteristic curves. Recalling Eqs. (4.198), (4.199) and (4.200), the output power $W = FV$ of the hydraulic cylinder/load servo system is defined as follows:

$$W = \eta_c \eta_v C I_R p_\ell \sqrt{p_s - p_\ell}. \tag{4.203}$$

Replacing differentiation dW/dp_ℓ by zero, the load pressure $p_\ell = p_p$ under the maximum power conditions is obtained as

$$p_\ell = p_p = \frac{2p_s}{3}. \tag{4.204}$$

Substituting Eq. (4.204) into Eqs. (4.199) and (4.200), the piston thrust F_p and the piston speed V_p under the maximum power conditions are obtained as follows:

$$F_p = \eta_c A p_p = \frac{2}{3} \eta_c A p_s, \tag{4.205}$$

$$V_p = \frac{\eta_v C I_R \sqrt{p_s - p_p}}{A} = \frac{\eta_v q_0}{\sqrt{3} A}. \tag{4.206}$$

Substituting the maximum flow rate $q_s = q_0$ shown in Eq. (4.202) into Eq. (4.199) gives the maximum piston speed V_m of the hydraulic system:

$$V_m = \frac{\eta_v q_0}{A} = \frac{\eta_v C I_R \sqrt{p_s}}{A}. \tag{4.207}$$

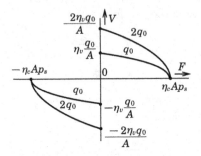

(a) With flow rate varying between q_0 and $2q_0$

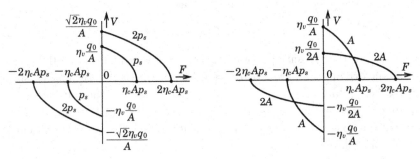

(b) With supply pressure varying
between p_s and $2p_s$

(c) With piston area varying
between A and $2A$

Fig. 4.50   Driving characteristic curves.

Substituting $p_\ell = p_s$ into Eq. (4.200) gives the maximum thrust

$$F_m = \eta_c A p_s. \tag{4.208}$$

4.4.1.3 *Design of a Hydraulic Servo System*

If a driving characteristic curve encloses the load chart as shown in Fig. 4.51, then the hydraulic system satisfies the power requirements implied by the load chart.

The hydraulic source power being $p_s q_0$, the hydraulic servo system efficiency η is given as

$$\eta = \frac{FV}{p_s q_0} = \eta_v \eta_c \frac{FV}{F_m V_m}. \tag{4.209}$$

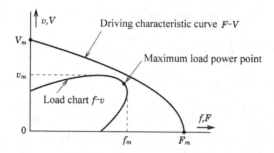

Fig. 4.51 Driving characteristic curve and a load chart.

Recalling Eqs. (4.205) and (4.206), the maximum power $F_p V_p$ in the driving characteristic curve is expressed as

$$F_p V_p = \frac{2\eta_c \eta_v p_s q_o}{3\sqrt{3}}. \tag{4.210}$$

Substituting $FV = F_p V_p$ into Eq. (4.209), the efficiency at the maximum power is obtained as

$$\eta = \frac{2}{3\sqrt{3}}\eta_v \eta_c = 0.385\eta_c \eta_v. \tag{4.211}$$

Equations (4.212) and (4.213) imply that the maximum power point in the driving characteristic curve should agree with that in the load chart

$$V_p = v_p, \tag{4.212}$$

$$F_p = f_p. \tag{4.213}$$

The condition may be used as a criterion to determine the servo valve capacity and actuator dimension which will yield a driving characteristic curve that fully encircles the load chart.

The piston area $A = A_a$ satisfying the condition given in Eq. (4.212) is obtained by substituting the Eqs. (4.193) and (4.206) into Eq. (4.212)

$$A_a = \frac{\eta_v q_0}{\sqrt{3}a\omega_b \cos(\phi/2)}. \tag{4.214}$$

No-load flow rate $q_0 = q_{0m}$ which should satisfy the condition given by Eqs. (4.212) and (4.213) is derived by canceling the parameters f_p, F_p and

$A \equiv A_a$ in Eqs. (4.194), (4.205), (4.213) and (4.214)

$$q_{0m} = \frac{3\sqrt{3}a^2\omega_b^2}{4\eta_v\eta_c p_s} \left[\left(m\omega_b - \frac{k}{\omega_b} \right) \sin\phi + 2B\cos^2\frac{\phi}{2} \right]. \tag{4.215}$$

Since the hydraulic servo system should be designed so that the driving characteristics curve encloses the load chart with a little margin, the no-load flow rate $q_0 = \delta q_{0m}$ should be required at the servo valve capacity slightly larger than q_{0m}. The condition whereby the driving characteristics curve should be given with the same margin against the maximum load force and the maximum load velocity is expressed as follows:

$$\frac{F_m}{f_m} = \frac{V_m}{v_m}, \tag{4.216}$$

$$q_0 = \delta q_{0m}. \tag{4.217}$$

Recalling Eqs. (4.196), (4.197), (4.207) and (4.208), the piston area $A = A_\beta$ satisfying the Eq. (4.216) is derived as follows:

$$A_\beta = \sqrt{\frac{\eta_v q_0}{\eta_c\omega_b p_s}} \left[\frac{f_m}{a} \right]^{1/2} = \sqrt{\frac{\eta_v q_0}{\eta_c\omega_b p_s}} [(m\omega_b^2 - k)^2 + (B\omega_b)^2]^{1/4}. \tag{4.218}$$

Since $A_\alpha = A_\beta]_{q_0=\delta q_{0m}}$ gives $\delta = 2/\sqrt{3}$, then recalling Eqs. (4.215) and (4.217), $q_0 = \delta q_{0m}$ is described as

$$q_0 = \frac{2}{\sqrt{3}}q_{0m} \approx 1.16\, q_{0m}. \tag{4.219}$$

Equations (4.214), (4.215), (4.218) and (4.219) provide us a valuable insight into the design of hydraulic servo system such as to satisfy the required response speed.

4.4.1.4 Design Procedure

The dimensions of a hydraulic servo system are designed in accordance with the design procedure (1) to (3), taking into account the results summarized in the previous section.

(1) Assume an adequate supply pressure $p_s = p_{s1}$. For this assumed supply pressure $p_s = p_{s1}$, the no-load flow rate q_{0m} may be estimated from Eq. (4.215) since the value of ϕ can be obtained from Eq. (4.192) using the values of parameters m, ω_b, k, B provided in the design specifications.

(2) The no-load flow rate $q_0 = \delta q_{0m}$ of the servo valve is obtained using the expression $\delta = 1.16$ given in Eq. (4.219) and q_{0m} estimated in the previous section. For the assumed supply pressure $p_s = p_{s1}$, the piston area $A = A_{\alpha 1}$ is estimated using Eq. (4.214).

(3) When the supply pressure $p_s = p_{s1}$ is altered to $p_s = p_s^*$, the no-load flow rate $q_0 = q_0^*$ and the piston area $A = A^*$ vary in accordance with the formulas derived by Eqs. (4.214) and (4.215):

$$q_0^* = \frac{p_{s1}}{p_s^*} q_0, \quad A^* = \frac{p_{s1}}{p_s^*} A_\alpha. \tag{4.220}$$

Therefore, the values of p_s, A and q_0 can be easily reselected in accordance with Eq. (4.220).

[Example 4.1]

In the hydraulic servo system illustrated in Fig. 4.47, the load conditions are given as follows. The load mass is $m = 180\,\mathrm{kg}$, the coefficient of viscous resistance force is $B = 3600\,\mathrm{Ns/m}$ and the spring force is regarded as zero ($k = 0$). It is required that the equivalent time constant in the system response is $T_e = 0.05\,\mathrm{s}$ for the piston displacement $a = 20\,\mathrm{mm}$. Discuss the main dimensions of the hydraulic servo system assuming that the thrust efficiency and volumetric efficiency take the value unity ($\eta_v = \eta_c = 1$).

[Solution 4.1]

The bandwidth ω_b needed for the servo system is $\omega_b = 1/T_e = 20\,\mathrm{s}^{-1}$. Therefore, $\phi = \tan^{-1}(180 \times 20/3600) = 45°$ is estimated using Eq. (4.192). Recalling Eq. (4.215), the no-load flow rate q_{0m} is estimated for the assumed supply pressure $p_s = 7\,\mathrm{MPa}$:

$$q_{0m} = \frac{3\sqrt{3} \times (4 \times 10^{-4}) \times 400}{4 \times (7 \times 10^6)} \times \left[\frac{180 \times 20}{\sqrt{2}} + 2 \times 3600 \times (\cos 22.5°)^2 \right]$$
$$= 258 \times 10^{-6}\,\mathrm{m^3/s}. \tag{1}$$

Recalling Eqs. (1) and (4.219), no-load flow rate $q_0 = \delta q_{0m}$ is obtained as follows:

$$q_0 = 1.16 \times (258 \times 10^{-6}) = 299 \times 10^{-6}\,\mathrm{m^3/s}. \tag{2}$$

Substituting $p_s = 7$ MPa and $q_0 = 299 \times 10^{-6}\,\mathrm{m^3/s}$ into Eq. (4.202) yields CI_R as

$$CI_R = \frac{299 \times 10^{-6}}{\sqrt{7 \times 10^6}} = 0.113 \times 10^{-6}\,\mathrm{m^3/(s \cdot Pa^{1/2})}. \tag{3}$$

Equation (4.214) gives the piston area $A = A_\alpha$

$$A_\alpha = \frac{299 \times 10^{-6}}{\sqrt{3} \times (2 \times 10^{-2}) \times 20 \times \cos 22.5^\circ} = 4.67 \times 10^{-4}\,\mathrm{m^2}. \tag{4}$$

For the supply pressure $p_s = 7\,\mathrm{MPa}$, the piston area $A = A_\alpha = 4.67 \times 10^{-4}\,\mathrm{m^2}$ and servo valve capacity $CI_R = 0.113 \times 10^{-6}\,\mathrm{m^3/(sPa^{1/2})}$, the driving characteristic equation corresponding to Eq. (4.201) is denoted as follows:

$$V = 0.0112 \times \sqrt{3270 - F}. \tag{5}$$

When the supply pressure $p_s = 7\,\mathrm{MPa}$ is changed to $p_s = 10\,\mathrm{MPa}$, the no-load flow rate q_0, servo valve capacity CI_R and piston area $A = A_\alpha$ can be derived from Eqs. (4.202) and (4.220):

$$q_0 = \frac{7}{10} \times (299 \times 10^{-6}) = 209 \times 10^{-6}\,\mathrm{m^3/s}, \tag{6}$$

$$CI_R = \frac{q_0}{\sqrt{p_s}} = \frac{209 \times 10^{-6}}{\sqrt{10 \times 10^6}} = 0.662 \times 10^{-7}\,\mathrm{m^3/(s \cdot Pa^{1/2})}, \tag{7}$$

$$A = \frac{7}{10} \times (4.67 \times 10^{-4}) = 3.27 \times 10^{-4}\,\mathrm{m^2}. \tag{8}$$

Equation (5) gives not only the driving characteristics for the combination of $p_s = 7\,\mathrm{MPa}$, $CI_R = 0.113 \times 10^{-6}\,\mathrm{m^3/(Pa^{1/2}s)}$ and $A = A_\alpha = 4.67 \times 10^{-4}\,\mathrm{m^2}$ but also the driving characteristics for $p_s = 10\,\mathrm{MPa}$, $CI_R = 0.662 \times 10^{-7}\,\mathrm{m^3/(Pa^{1/2}s)}$ and $A = A_\alpha = 3.27 \times 10^{-4}\,\mathrm{m^2}$.

The combinations of supply pressure p_s, servo valve capacity CI_R and piston area A should be selected carefully, in consideration of the specificity of the application. Substituting the design specifications given in the example problem into Eq. (4.187) gives the values of coefficients α, β and

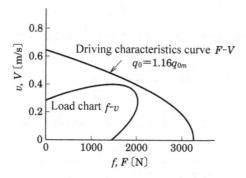

Fig. 4.52 Driving characteristic curve enclosing the load chart.

γ in Eq. (4.186)

$$\left.\begin{array}{l} \alpha = 3.86 \times 10^{-8}\,\mathrm{m^2/(Ns)^2} \\ \beta = 1.39 \times 10^{-4}\,\mathrm{m/(Ns)} \\ \gamma = 8 \times 10^{-2}\,\mathrm{m^2/s^2} \end{array}\right\}. \qquad (9)$$

Thus, Eq. (4.186) and Eq. (9) yield the load chart of an ellipse with the major axis $a^* = 2330$ N, the minor axis $b* = 0.247\,\mathrm{m/s}$ and inclined angle $\theta = 31.7°$ for force-velocity scale ratio $n = 1/3600$. It can be readily seen in Fig 4.52 that the driving characteristic curve of Eq. (5) effectively encloses the load chart given by Eq. (4.186) and Eq. (9).

4.4.2 *Control Performance of a Hydraulic Servo System*

4.4.2.1 *Synthesis of Control Performance*

It is possible to synthesize the control performance of a servo system provided that the open loop dynamic characteristics are given as the approximated transfer function. In this section, the synthesis of the control performance using the frequency method will be considered. Figure 4.53 shows the block diagram of the hydraulic servo cylinder system. The sensor transfer functions $G_t(s)$ and $G_R(s)$ have to have the identical gain constant K_t in order to ensure that the output displacement Δy agrees with the command Δy_R, since the response speed is fast enough to be ignored

$$G_R(s) = G_t(s) = K_t. \qquad (4.221)$$

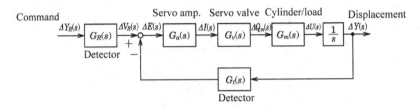

Fig. 4.53   Hydraulic servo cylinder system.

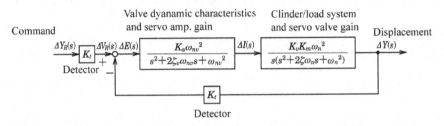

Fig. 4.54   Block diagram of a hydraulic servo cylinder system.

As explained in Section 4.3.1, the transfer function of a servo amplifier without a compensator can be written as

$$G_a(s) = K_a. \tag{4.222}$$

The diagrammatic representation of a system such as that shown in Fig. 4.53 can be simplified to the block diagram shown in Fig. 4.54. This is because each transfer function of the servo valve and the cylinder/load system is defined by a standard second order delay system, as discussed in Sections 4.3.2 and 4.3.3. However, in the neighborhood of the gain cross-over frequency, the Bode diagram of the open loop system is hardly affected by the servo valve characteristics when the undamped natural frequencies ω_{nv} of the servo valve exceed the ten-fold value of the undamped natural frequencies ω_n of the cylinder/load system:

$$\omega_{nv} > 10\omega_n. \tag{4.223}$$

Thus, Fig. 4.54 is arranged as Fig. 4.55, and the open loop transfer function $K\overline{G}(s)$ is given as follows:

$$\left. \begin{aligned} K\overline{G}(s) &= G_1(s)G_2(s) \\ G_1(s) &= \frac{K_t K_a K_v K_m}{s} = \frac{K_0}{s}, \quad G_2(s) = \frac{\omega_n^2}{s^2 + 2\zeta\omega_n s + \omega_n^2} \end{aligned} \right\}. \tag{4.224}$$

Command   Displacement

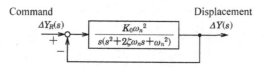

Fig. 4.55   Simplified block diagram of a hydraulic servo cylinder system.

When the undamped natural angular frequency ω_n exceeds the five-fold value of the open loop gain constant K_0, the magnitude of $K\overline{G}(j\omega) = G_1(j\omega)G_2(j\omega)$ may be approximated by the magnitude of $G_1(j\omega) = K_0/(j\omega)$ in the neighborhood of the gain crossover frequency on the Bode diagram:

$$\omega_n > 5K_0. \tag{4.225}$$

Then gain crossover frequency ω_g is given as follows:

$$\omega_g = K_0. \tag{4.226}$$

Let us consider an example: the open loop transfer function $K\overline{G}(s) = K_0\omega_n^2/[s(s^2 + 2\zeta\omega_n s + \omega_n^2)]$ which involves $G_1(s) = K_0/s$ with $K_0 = 31.4\,\text{rad/s}$ and $G_2(s) = \omega_n^2/(s^2 + 2\zeta\omega_n s + \omega_n^2)$ with $\zeta = 0.33$, $\omega_n = 300\,\text{rad/s}$. As shown in Fig. 4.56, the gain crossover frequency of $G_1(j\omega) = K_0/(j\omega)$ is written as $\omega_g = K_0 = 31.4\,\text{rad/s}$. Then the logarithmic magnitude of $G_2(j\omega_g)$ is regarded as nearly zero in the frequency range of $\omega \leq 1.5\omega_g$ by virtue of the condition $\omega_n \geq 5K_0$.

Therefore, the magnitude $|K\overline{G}(j\omega)|$ can be approximated by $|G_1(j\omega)| = |K_0/(j\omega)|$ for $\omega \leq 1.5\omega_g$ as shown in Fig. 4.56. Recalling Eqs. (4.61) and (4.64), the phase margin ϕ_m of the hydraulic servo system is given as

$$\phi_m = 90° + \angle G_2(j\omega_g) = 90° - \tan^{-1}\frac{2\zeta(\omega_g/\omega_n)}{1 - (\omega_g/\omega_n)^2}. \tag{4.227}$$

Since the gain crossover frequency ω_g depends on the loop gain $K_0 = K_t K_a K_v K_m$, it can be adjusted by altering the servo amplifier gain K_a so as to give a desirable phase margin ϕ_m. When the phase margin ϕ_m is larger than about 70° to 80°, the dynamic behavior of the system can be approximated by a first order lag system with an equivalent time constant

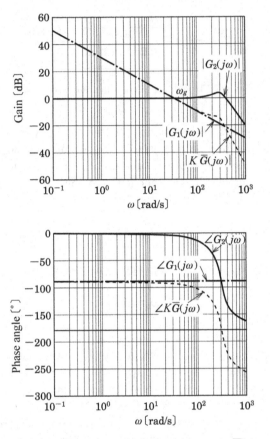

Fig. 4.56   Bode diagram of an open loop transfer function $K\overline{G}(s) = G_1(s)G_2(s)$.

T_e defined by the following equation:

$$T_e \approx \frac{1}{\omega_g} = \frac{1}{K_0}. \qquad (4.228)$$

Therefore, the gain crossover frequency ω_g can be used as a reference to estimate the response speed of the control system.

When the condition implied by Eq. (4.225) is not satisfied, the synthesis of the servo system control performance should involve the open loop transfer function $K\overline{G}(s) = G_1(s)G_2(s)$ given by Eq. (4.224). The Bode diagram for Eq. (4.224) obtained by adding graphically both Bode diagrams of $G_1(j\omega)$ and $G_2(j\omega)$ is denoted by the dotted line in Fig. 4.56. It is readily apparent in the Bode diagram that the phase margin ϕ_m tends to decrease as the crossover frequency ω_g approaches the natural angular frequency ω_n due to the increase of the gain constant K_0. In other words, the increase of the open loop gain K_0 causes the stability of the servo system to deteriorate, at the same time enhancing the response speed.

The countermeasure which can be employed in the synthesis of control performance, involving the series compensation to improve the stability without compromising the response speed and the control accuracy will be discussed in Section 4.4.3. For the hydraulic servo system failing to satisfy the condition given in Eq. (4.223), the control performance should be discussed taking into account the open loop transfer function $K\overline{G}(s)$ as shown in the following equation:

$$K\overline{G}(s) = \frac{K_0 \omega_{nv}^2 \omega_n^2}{s(s^2 + 2\zeta_v \omega_{nv} s + \omega_{nv}^2)(s^2 + 2\zeta_n \omega_n s + \omega_n^2)}. \qquad (4.229)$$

4.4.2.2 Control Accuracy in the Steady State

The steady state position error of the hydraulic servo system is theoretically zero because it is a one-type control system with an integral element. However, in practice, the steady state position is in error due to the drift in the servo valve system. Although the drift depends on the small fluctuation of the spool position due to servo amplifier noise, spool Coulomb friction and mechanical backlash in the servo system, etc., it may be treated as the equivalent electrical current disturbance Δi_d given at the amplifier output as shown in Fig. 4.57. The magnitude of Δi_d is in the range from one percent to three percent of the rated maximum electrical current I_R of the servo amplifier.

Assuming the command value $\Delta Y_R(s)$ to be zero in Fig. 4.57, the transfer function $\Delta Y(s)/\Delta I_d(s)$ is obtained by simplifying the block diagram:

$$\frac{\Delta Y(s)}{\Delta I_d(s)} = \frac{K_v K_m}{s^3 + 2\zeta \omega_n s^2 + \omega_n^2 s + K_0}, \qquad (4.230)$$

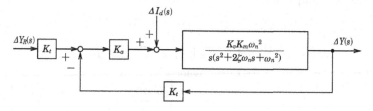

Fig. 4.57 Hydraulic servo system involving a disturbance.

where, the loop gain K_0 is defined as $K_0 = K_t K_a K_v K_m$ shown in Fig. 4.54. Recalling Table 4.2(m), the steady position error ε_p (steady position accuracy) is obtained for the applied disturbance Δi_d:

$$\varepsilon_p = \lim_{s \to 0} s\Delta Y(s) = \frac{\Delta i_d}{K_t K_a} = \frac{\Delta i_d}{I_R} \frac{I_R K_v K_m}{K_0} = \frac{\Delta i_d}{I_R} \frac{V_m}{K_0}, \qquad (4.231)$$

where $V_m = I_R K_v K_m$ stands for the maximum steady-state piston speed.

4.4.3 *Compensation Techniques*

Before discussing the control performance of the servo system, one needs to determine the servo valve capacity, the actuator dimensions and the supply pressure that will provide the required driving power. This may be done using the procedure discussed in Section 4.4.1. The control performance is established using the assumption that the servo system is sufficiently powerful to drive the load with the specified response speed.

With regard to the target hydraulic servo system, the mathematical model of the servo system should be first estimated theoretically and experimentally. Then, the stability of the servo system with the loop gain K_0 satisfying the steady accuracy and the response speed condition should be confirmed. This is done by investigating the phase margin of the open loop transfer function on the Bode diagram. Since the gain adjustment which should satisfy the response speed and the steady state accuracy frequently provides poor stability, a compensation strategy should be adopted to improve the control performances. This is done by connecting a suitable electrical network, called a compensator, to the servo amplifier. Figure 4.58 illustrates a series compensation method in which the compensator is connected in series with the open loop system. The phase lead compensator, the phase lag compensator, the phase lead-lag compensator and the PID controller are all widely used as series compensators, and the electrical networks of the compensators are shown in Table 4.8.

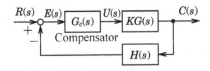

Fig. 4.58   Series compensation.

4.4.3.1 *Phase Lead, Lag and Lead-lag Compensation*

(a) *Phase lead compensation*

The transfer function $G_c(s)$ of the phase lead compensator is given as

$$G_c(s) = \frac{U(s)}{E(s)} = \frac{1 + aTs}{1 + Ts} \quad (a > 1). \tag{4.232}$$

Equation (4.232) gives

$$\lim_{s \to 0} G_c(s) = 1. \tag{4.233}$$

Equation (4.233) implies that the phase lead compensator does not change the steady performances as given by Eqs. (4.42), (4.43) and (4.45) in the case of the steady errors discussed in Section 4.2.3. Recalling Eq. (4.232), the magnitude $|G_c(j\omega)|$ and phase angle $\angle G_c(j\omega)$ are expressed as

$$20 \log |G_c(j\omega)| = 10\{\log[1 + (aT\omega)^2] - \log[1 + (T\omega)^2]\} \quad \text{[dB]}, \tag{4.234}$$

$$\angle G_c(j\omega) = \tan^{-1}(aT\omega) - \tan^{-1}(T\omega) = \tan^{-1} \frac{(a-1)T\omega}{1 + a(T\omega)^2}. \tag{4.235}$$

The angular frequency ω_m at the maximum phase angle $\phi_m = \angle G_c(j\omega_m)$ is derived by rearranging $d[\angle G_c(j\omega)]/d\omega = 0$:

$$\omega_m = \frac{1}{T\sqrt{a}}. \tag{4.236}$$

The maximum phase angle $\phi_m = \angle G_c(j\omega_m)$ and magnitude $|G_c(j\omega_m)|$ at $\omega = \omega_m$ are expressed as

$$\phi_m = \angle G_c(j\omega_m) = \sin^{-1} \frac{a-1}{a+1}, \tag{4.237}$$

$$20 \log |G_c(j\omega_m)| = 10 \log a \quad \text{[dB]}. \tag{4.238}$$

Figure 4.59 illustrates the Bode diagram of $G_c(j\omega)$ in which the broken line denotes the straight-line approximation of the gain curve with the corner frequencies $\omega = 1/(aT)$ and $\omega = 1/T$.

Let us now consider a servo system with an open loop transfer function $K\overline{G}(s) = KG(s)H(s)$ as shown in Fig. 4.60(a). The Bode diagram of $K\overline{G}(j\omega) = KG(j\omega)H(j\omega)$ is plotted with solid line in Fig. 4.61, and it is assumed that the phase margin ϕ_1 should be compensated to obtain the desired stability, though both the response speed and the steady control accuracy are desirable. Thus, the phase lead compensator governed by the

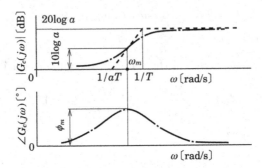

Fig. 4.59   Bode diagram of a phase lead compensator.

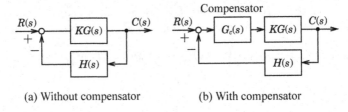

(a) Without compensator   (b) With compensator

Fig. 4.60   Servo systems without and with compensator.

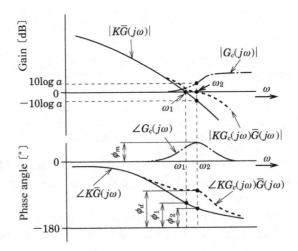

Fig. 4.61   Phase lead compensation technique.

transfer function $G_c(s)$ is set in series in the open loop control system in order to compensate the phase margin, as shown in Fig. 4.60(b).

Recalling that the Bode diagram of the compensated open loop transfer function $K\overline{G}(s)G_c(s)$ may be depicted by the graphic addition of $G_c(j\omega)$ to $K\overline{G}(j\omega)$, it is evident that the phase margin of $K\overline{G}(j\omega)G_c(j\omega)$ is most effectively compensated when the factor a and T are selected so as to make the gain crossover frequency ω_2 of $K\overline{G}(j\omega)G_c(j\omega)$ coincident with angular frequency ω_m at the maximum phase angle $\phi_m = \angle G_c(j\omega_m)$. Figure 4.61 illustrates the Bode diagram of $K\overline{G}(j\omega)G_c(j\omega)$, the broken lines denoting the graphically added Bode diagram of $G_c(j\omega)$ to the Bode diagram of $K\overline{G}(j\omega)$. Consequently, the gain crossover frequency moves to the frequency ω_2, and then the phase margin increases to the desired value ϕ_d due to the phase lead compensator. Therefore, the stability and response speed of the control system are improved by the phase lead compensator without causing the steady-state performance to deteriorate.

The procedure employed to estimate the phase lead compensator factors a, T is given as follows.

(1) Depict the Bode diagram of the open loop transfer function $K\overline{G}(s)$ satisfying both the response speed and the steady accuracy conditions given in the specifications and find the phase margin ϕ_1 accordingly. If phase margin ϕ_1 is less than the desired value, the phase angle ϕ_0 to compensate the phase margin is derived from the following equation:

$$\phi_0 = \phi_d - \phi_1 + \alpha, \tag{4.239}$$

where, ϕ_d is the desired phase margin, and α is a factor to make up phase angle change $\angle K\overline{G}(j\omega_2) - \angle K\overline{G}(j\omega_1)$ in accordance with the crossover frequency change. Phase angle ϕ_0 is temporarily decided by adopting a trial value of α, which generally lies in the appropriate range. Substituting $\phi_m = \phi_0$ into Eq. (4.237) gives parameter a in $G_c(s)$.

(2) Find the gain crossover frequency ω_2 which can be obtained as the frequency at $|K\overline{G}(j\omega_2)| = -10\log a$ on the Bode diagram:

$$20\log|KG(j\omega_2)| = -10\log a \quad [\text{dB}]. \tag{4.240}$$

Substituting the values of $\omega_2 = \omega_m$ and a into Eq. (4.236), the factor T in $G_c(s)$ is expressed as

$$T = \frac{1}{\omega_2\sqrt{a}}. \tag{4.241}$$

(3) It should be confirmed by depicting the Bode diagram of $KG_c(j\omega)\overline{G}(j\omega)$ whether the value of the phase margin is desirable or not.

Unless the phase margin reaches the desired value, the procedures from steps (1) to (3) should be repeated whilst the value of the correction factor α should be increased.

(b) *Phase lag compensation*

Though the phase lag compensator transfer function $G_c(s)$ is also defined by Eq. (4.232), it is different from a phase lead compensator in that the value of factor a is positive and less than one:

$$G_c(s) = \frac{1 + aTs}{1 + Ts} \quad (0 < a < 1). \tag{4.242}$$

Therefore the magnitude $|G_c(j\omega)|$ and phase $\angle G_c(j\omega)$ of a phase lag compensator are expressed by Eqs. (4.234) and (4.235), respectively and the minimum phase angle $\phi_m = \angle G_c(j\omega_m)$ and frequency ω_m are given by Eqs. (4.237) and (4.236). Thus, both the logarithmic magnitude $20 \log |G_c(j\omega)|$ and the phase angle $\angle G_c(j\omega)$ take a minus value because of the condition $0 < a < 1$. Figure 4.62 illustrates the Bode diagram of a phase lag compensator, and Fig. 4.63 compares the Bode diagram of $K\overline{G}(j\omega)$ with the Bode diagram of $KG_c(j\omega)\overline{G}(j\omega)$.

Appending a phase lag compensator in series to open loop transfer function $K\overline{G}(s)$, the magnitude of $K\overline{G}(j\omega)G_c(j\omega)$ is reduced to $20 \log |K\overline{G}(j\omega)| + 20 \log a$ for frequency range $\omega \gg 1/(aT)$, and simultaneously the crossover frequency transfers to ω_2.

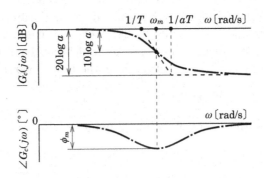

Fig. 4.62   Bode diagram of a phase lag compensator.

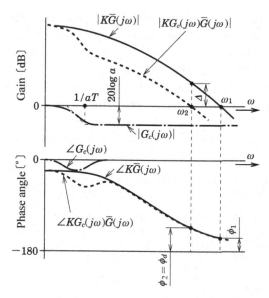

Fig. 4.63   Lag compensation technique.

Thus, the phase angle $\angle KG_c(j\omega)\overline{G}(j\omega)$ becomes nearly equal to $\angle K\overline{G}(j\omega)$ in the neighborhood of the crossover frequency ω_2 provided that parameters a and T are selected so as to satisfy the following Eq. (4.243):

$$\frac{1}{aT} \approx \frac{\omega_2}{10}. \tag{4.243}$$

Thereby, the phase margin increases to ϕ_2 due to decrease of crossover frequency, because the phase angle $\angle K\overline{G}(j\omega)$ generally decreases with the increase of frequency ω. Though a phase lag compensator does not affect the steady error performances as well as a phase lead compensator, it improves the stability of the control system, at the same time slowing the response speed through the decrease of the crossover frequency.

The procedure to estimate phase lag compensator factors a and T is outlined as follows.

(1) First select the open loop gain constant K to ensure the desired steady–state performances, in accordance with Eq. (4.231) and Table 4.6 in Section 4.2.3.

(2) Find the phase margin ϕ_1 and crossover frequency ω_1 for the uncompensated control system by plotting the Bode diagram of $K\overline{G}(j\omega)$ as shown in Fig. 4.63. If the phase margin ϕ_1 is less than the required value,

the crossover frequency ω_1 should be transferred to the frequency ω_2 to get the desired phase margin $\phi_2 = \phi_d$ by decreasing the magnitude of $K\overline{G}(j\omega)$ to $(20\log|K\overline{G}(j\omega)| - \Delta)$ dB in the frequency range where $\angle G_c(j\omega)$ hardly affects the $\angle G_c(j\omega)\overline{G}(j\omega)$.

Reading off the magnitude of $\Delta = |K\overline{G}(j\omega_2)|$ dB from the Bode diagram at the crossover frequency ω_2, the factor a is obtained recalling Fig. 4.62:

$$-\Delta = 20\log a \ [\text{dB}]. \qquad (4.244)$$

Substituting the values of ω_2 and a into Eq. (2.243) yields the value of T.

(3) Verify the value of the phase margin for the compensated control system by graphically superimposing $G_c(j\omega)$ and $K\overline{G}(j\omega)$ on the Bode diagram.

(c) *Lead-lag compensation*

Not only does a phase lead compensation improve the system stability but it also promotes the response speed. The lag compensation improves stability without degrading the steady-state performance, although it does slow down the response speed.

Since the phase margin compensation is shared between the phase lead and the lag compensation, the phase lead-lag compensator provides the desired phase margin without changing other desired control performances. The transfer function of a phase lead-lag compensator is defined as

$$G_c(s) = G_{c1}(s)G_{c2}(s)$$

$$\left. G_{c1}(s) = \frac{1 + a_1 T_1 s}{1 + T_1 s} \quad (a_1 > 1), \quad G_{c2}(s) = \frac{1 + a_2 T_2 s}{1 + T_2 s} \quad (0 < a_2 < 1) \right\}.$$

$$(4.245)$$

An example of the phase lead-lag compensation is discussed in Section 4.4.4.

4.4.3.2 *Compensation by Means of a PID Controller*

(a) *PID controller*

A PID controller is often used as a compensator which provides a proportional control action, an integral control action and a derivative control action for the control error input $e(t)$. These are referred to as P action, I action and D action, respectively

$$u(t) = K_p e(t) + K_i \int e(t)dt + K_d \frac{de(t)}{dt}. \qquad (4.246)$$

Where K_p, K_i and K_d are proportional, integral and derivative gains respectively.

The Laplace transformation of Eq. (4.246) gives the PID controller transfer function $G_c(s)$:

$$G_c(s) = \frac{U(s)}{E(s)} = K_p \left(1 + \frac{1}{T_i s} + T_d s \right), \qquad (4.247)$$

where factors T_i and T_d are called an integral time and a derivative time respectively, and K_p may be regarded as an adjusting gain of the open loop transfer function in the control system. Figure 4.64 illustrates the block diagram of a PID controller.

Let us consider a PI controller capable of PI action. Assuming the derivative time T_d to be zero in Eq. (4.247), the PI controller transfer function $G_c(s)$ is expressed as follows:

$$G_c(s) = K_p G_{\mathrm{PI}}(s), \quad G_{\mathrm{PI}}(s) = 1 + \frac{1}{T_i s}. \qquad (4.248)$$

Figure 4.65 shows the Bode diagram of $G_{\mathrm{PI}}(j\omega)$. As shown in Fig. 4.65, the frequency transfer function $G_{\mathrm{PI}}(j\omega)$ is approximated by $1/(j\omega T_i)$ for the low frequency range $\omega < 0.1/T_i$ and by $G_{\mathrm{PI}}(j\omega) = 1$ for the high frequency range $\omega > 10/T_i$, thus, the controller transfer function $K_P G_{\mathrm{PI}}(j\omega)$ hardly affects the stability of the servo system provided that gain crossover frequency ω_g in $K\overline{G}(j\omega)K_p G_{\mathrm{PI}}(j\omega)$ is larger than $10/T_i$.

PID controller $G_c(s)$

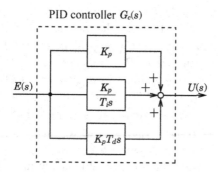

Fig. 4.64   PID controller.

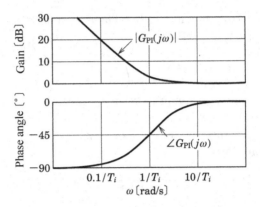

Fig. 4.65   Bode diagram of $G_{PI}(s) = 1 + 1/(T_i s)$.

Since $G_{PI}(s)$ is approximated by $1/(T_i s)$ at infinitesimal values of the complex variable s, the PI controller transfer function $K_p G_{PI}(s)$ leads to an increase of the exponent q of the Laplace operator s in the open loop transfer function as discussed in Eq. (4.44) so that it acts as a compensator to improve the steady performance regardless of the value of K_p. Moreover, the stability of the control system is improved by decreasing the proportional gain K_p, although it causes the response speed to slow down. It is reasonable to suppose, therefore, that a PI controller $G_c(s) = K_p G_{PI}(s)$ acts as a lag compensator provided that factors K_p and T_i are suitably chosen.

Assuming the integral time to be $T_i = \infty$ in Eq. (4.247), the PD controller transfer function $G_c(s)$ is expressed as

$$\left.\begin{array}{l} G_c(s) = K_p G_{PD}(s) \\ G_{PD}(s) = 1 + T_d s \end{array}\right\}. \qquad (4.249)$$

Figure 4.66 shows the Bode diagram of $G_{PD}(s) = 1 + T_d s$. The magnitude of $G_{PD}(j\omega) = 1 + jT_d\omega$ in the Bode diagram is nearly zero dB for the frequency range $\omega < 1/T_d$, and then it approximates the line with the slope 20 dB/decade passing through 0 dB at frequency $\omega = 1/T_d$ for the frequency range $\omega > 1/T_d$. The phase angle $\angle G_{PD}(j\omega)$ increases from about zero to $90°$ in the frequency range $0.1/T_d \le \omega \le 10/T_d$ though it is nearly zero in the range $\omega < 0.1/T_d$. Therefore, it is supposed that the PD transfer function $G_{PD}(j\omega) = K_p(1 + T_d s)$ acts as a phase lead compensator discussed previously. For instance, selecting factors T_d and K_p so as to make factor $1/T_d$ coincide with the crossover frequency in $K_p G_{PD}(j\omega)K\overline{G}(j\omega)$, the phase margin increase of about $45°$ may be attained, and then the

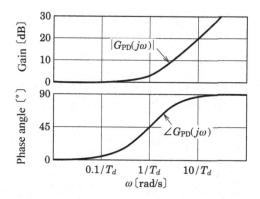

Fig. 4.66　Bode diagram of $G_{\text{PD}}(s) = 1 + T_d s$.

response speed is also enhanced provided that the crossover frequency in $K_p G_{\text{PD}}(j\omega) K\overline{G}(j\omega)$ is a little larger than in $K\overline{G}(j\omega)$.

In practical applications, the PD-controller transfer function $K_p(1 + T_d s)/(1 + 0.1 T_d s)$ is used instead of $K_p(1 + T_d s)$ so that the high frequency noises in the control signal are eliminated by the first order lag element $1/(1 + 0.1 T_d)$.

The product of $G_{\text{PI}}(s)$ by $G_{\text{PD}}(s)$ can be expressed in the same form in as Eq. (4.247), as shown in Eq. (4.250)

$$G_{\text{PID}}(s) = G_{\text{PI}}(s)G_{\text{PD}}(s) = \frac{(1 + T_i s)(1 + T_d s)}{T_i s}$$

$$= \frac{T_i + T_d}{T_i}\left[1 + \frac{1}{(T_i + T_d)s} + \frac{T_i T_d}{(T_i + T_d)}s\right]. \qquad (4.250)$$

It is reasonable to suppose, therefore, that the PID controller acts as a phase lead-lag compensator.

(b) *Tuning approach of a PID controller*

PID controller factors K_p, T_i and T_d in Eq. (4.247) are determined by the experimental approach based on Ziegler–Nichols rules. They are most useful in practical applications because mathematical models of hydraulic servo systems are not easy to derive.

The PID controller factors being selected in accordance with the Ziegler–Nichols rules, the indicial response of the compensated control system takes the overshoot from 0.1 to 0.6 ordinarily. Hence it is necessary to further adjust the controller factors, and this is usually done by trial and error. Ziegler–Nichols rules involve either the first or second method given in the following descriptions.

(1) *First method in Ziegler–Nichols rules*

When the system exhibits an indicial response with the S-shaped curve as depicted in Fig. 4.9, the transfer functions may be approximated as follows:

$$KG(s) = \frac{Ke^{-Ls}}{Ts+1}. \tag{4.251}$$

Pure dead time L and time constant T are obtained using the approach outlined in Section 4.2.2.

The first method in the Ziegler–Nichols rules is applicable to an open loop system denoted by Eq. (4.251), and it gives the factors of PID controller as follows:

PID controller:

$$\left.\begin{aligned} K_p = 1.2(T/L), \quad T_i = 2L, \quad T_d = 0.5L \\ G_c = 0.6T\frac{[s+(1/L)]^2}{s} \end{aligned}\right\}. \tag{4.252}$$

PI controller:

$$K_p = 0.9(T/L), \quad T_i = L/0.3, \quad T_d = 0. \tag{4.253}$$

P controller:

$$K_p = T/L, \quad T_i = \infty, \quad T_d = 0. \tag{4.254}$$

(2) *Second method in Ziegler–Nichols rules*

The second method in Ziegler–Nichols rules is applicable to a control system consisting of an open loop system with an integral element or an open loop system producing an oscillatory response, and the controller factors are given as:

PID controller:

$$\left.\begin{aligned} K_p = 0.6K_c, \quad T_i = 0.5T_c, \quad T_d = 0.125T_c \\ G_c(s) = 0.075K_cT_c\frac{(s+4/T_c)^2}{s} \end{aligned}\right\}. \tag{4.255}$$

PI controller:

$$K_p = 0.45K_c, \quad T_i = T_c/1.2, \quad T_d = 0. \tag{4.256}$$

P controller:

$$K_p = 0.5K_c, \quad T_i = \infty, \quad T_d = 0, \tag{4.257}$$

where K_c denotes the stability limit gain for the control system with a P controller, and T_c denotes the oscillation period at the stability limit. Increasing proportional gain K_p from zero to the stability limit K_c in the actual control system, both K_c and $T_c = 2\pi/\omega_c$ can be obtained experimentally.

Let us examine the effectiveness of the second method in Ziegler–Nichols rules using the control system whose open loop transfer function is given as follows:

$$KG(s) = \frac{1}{s(s+1)(s+4)}, \quad H(s) = 1. \tag{4.258}$$

The P controller transfer function being denoted as K_p, the characteristic equation for the control system is described as follows:

$$s^3 + 5s^2 + 4s + K_p = 0. \tag{4.259}$$

Hurwitz stability criterion in Section 4.2.5 gives the stability limit $K_p = K_c$ as follows:

$$K_c = 20. \tag{4.260}$$

The angular frequency ω_s at the stability limit is obtained by substituting $K_p = 20$ and $s = j\omega_s$ into Eq. (4.259) because the characteristic roots include the imaginary roots at stability limit, as discussed in Section 4.2.5

$$\left. \begin{array}{l} \omega_s = 2\,\text{rad/s} \\ T_c = 2\pi/\omega_s = \pi \end{array} \right\}. \tag{4.261}$$

Substituting Eqs. (4.260) and (4.261) into Eq. (4.255) gives the PID controller factors

$$K_p = 12, \quad T_i = 1.57, \quad T_d = 0.393, \tag{4.262}$$

$$G_c(s) = \frac{4.71(s+1.27)^2}{s}. \tag{4.263}$$

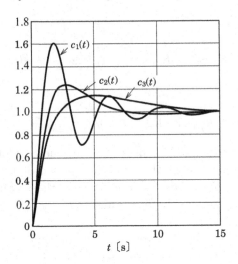

Fig. 4.67   Indicial responses for adjusted duplication zeros.

($K_p =12$, Duplication zeros $c_1(t)$: $s = -1.27$, $c_2(t)$: $s = -0.64$, $c_3(t)$: $s = -0.42$)

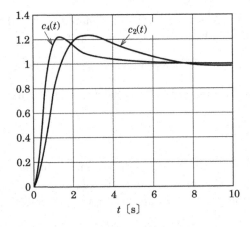

Fig. 4.68   Indicial responses for adjusted K_p.

(Duplication zeros $s = -0.64$, $c_2(t)$: $K_p = 12$, $c_4(t)$: $K_p = 24$)

Figures 4.67 and 4.68 show the simulation results on the indicial response obtained by computer methods. The curve $c_1(t)$ represents the indicial response of the control system with a PID controller setting the factors given in Eq. (4.262). The response curve $c_1(t)$ is a little unsatisfactory in terms of stability since the overshoot is about 0.6.

If duplication zeros $s = -1.27$ in Eq. (4.263) are altered to $s = -0.64$, then the indicial response for the proportional gain factor $K_p = 12$ changes to curve $c_2(t)$ while the overshoot is about 0.23. Altering the duplication zeros to $s = -0.42$, the response curve for $K_p = 12$ changes to $c_3(t)$ and the overshoot becomes about 0.15. When the proportional gain factor is altered from $K_p = 12$ to $K_p = 24$ for duplication zeros $s = -0.42$, the response curve becomes the curve $c_4(t)$ in Fig. 4.68. The curve $c_4(t)$ reveals that the response speed is improved without impairing stability in comparison with the response curve $c_2(t)$. In this manner, the controller factors determined in accordance with Ziegler–Nichols rules should be finely readjusted so that the control system exhibits satisfactory transient response.

4.4.4 *Design Example of a Hydraulic Servo System*

[**Example 4.2**]

Design a servo cylinder system driving the load mass $m = 500\,\text{kg}$ with the response speed of the equivalent time constant $T = 0.0333\,\text{s}$, for the displacement distance $a = \pm 10\,\text{mm}$. The design procedures are given as follows:

(1) The hydraulic servo cylinder system is shown in Fig. 4.69, and the control stability must be provided with the phase margin of 80° or more.
(2) A servo valve must be selected from options summarized in Table 4.10. The rated maximum output electric current I_R of the servo amplifier is 30 mA for rated maximum input voltage $E = 10\,\text{V}$ (i.e. gain $K_a = I_R/E = 3\,\text{mA/V}$).
(3) The response delays of the servo valve and the servo amplifier are negligible.
(4) The damping force, the spring force and the friction force are regarded as zero in the load system.

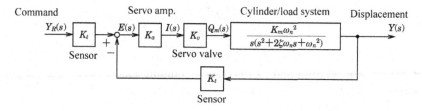

Fig. 4.69 Block diagram of a servo cylinder system.

Table 4.10   Servo valve characteristics.

Servo valve type	4.5L	7.5L	15L	25L	40L	60L
Rated pressure [MPa]			14			
Usable pressure [MPa]			1~21			
Rated flow rate [L/min]	4.5	7.5	15	25	40	60
Rated electric current [mA]			30			
Hysteresis		Less than 2% of rated electric current				
Frequency [Hz] at phase angle −90° or gain −6 dB for pressure 14 MPa and current input ±15 mA		100		80		50
Oil temperature [°C]			10~55			
Oil kinematic viscosity [cSt]			15~400			

(5) Both the thrust efficiency and the volumetric efficiency of the cylinder are assumed to have the value unity.

[Solution 4.2]

(1) *Power required to drive the load*

The power needed to move the load over the distance $y(\infty) = a$ at time constant $T = 1/\omega_b$ corresponds to the power needed to drive the load in the sinusoidal motion $y = a \sin \omega_b t$. Substituting $B = k = 0$ into Eq. (4.187) gives $\alpha = 1/(m\omega_b)^2$, $\beta = 0$ and $\gamma = (a\omega_b)^2$. Consequently, Eq. (4.186) is rearranged as Eq. (1) for the driving conditions

$$\frac{f^2}{(am\omega_b^2)^2} + \frac{v^2}{(a\omega_b)^2} = 1. \tag{1}$$

Equation (1) gives an elliptical load chart with the radius a^* and b^*,

$$a^* = f_m = am\omega_b^2 = 4500 \text{ N}, \tag{2}$$

$$b^* = v_m = a\omega_b = 0.30 \text{ m/s}. \tag{3}$$

Substituting $B = k = 0$ into Eq. (4.192) gives

$$\phi = \frac{\pi}{2}. \tag{4}$$

Recalling Eqs. (4.193) and (4.194), the maximum load power $w_p = f_p v_p$ is obtained as follows:

$$w_p = f_p v_p = 3180 \times 0.212 = 675 \text{ W} = 0.675 \text{ kW}. \tag{5}$$

(2) *Dimensions of hydraulic system elements*

A servo cylinder system should be designed so that the driving power curve should effectively enclose the load chart given in Eq. (1). Substituting $\phi = \pi/2$, $B = k = 0$ and $\eta_c = \eta_v = 1$ into Eq. (4.215) gives

$$q_{0m} = \frac{3\sqrt{3}a^2\omega_b^3 m}{4p_s}. \tag{6}$$

The maximum rated supply pressure p_{sm} of the servo valve is 14 MPa, as shown in Table 4.10. Adopting the supply pressure $p_s = 7$ MPa for the hydraulic servo system, the no-load flow rate q_{0m} corresponding to the maximum load power is estimated using Eq. (6)

$$q_{0m} = \frac{3\sqrt{3} \times (10 \times 10^{-3})^2 \times 30^3 \times 500}{4 \times (7 \times 10^6)}$$

$$= 0.251 \times 10^{-3} \, \mathrm{m}^3/\mathrm{s} = 15.1 \, \mathrm{L}/\min. \tag{7}$$

The required no-load flow rate q_{0r} is estimated recalling Eq. (4.219):

$$q_{0r} = 1.16 q_{0m} = 0.291 \times 10^{-3} \, \mathrm{m}^3/\mathrm{s} = 17.5 \, \mathrm{L}/\min. \tag{8}$$

Selecting the servo valve type 25L in Table 4.10, the no-load flow rate q_0 is estimated for the supply pressure $p_s = 7$ MPa recalling Eq. (4.202):

$$q_0 = 25 \times \sqrt{\frac{7}{14}} = 17.7 \, \mathrm{L}/\min = 0.295 \times 10^{-3} \, \mathrm{m}^3/\mathrm{s}. \tag{9}$$

The selected servo valve type 25L is suitable for the servo system since the no-load flow rate q_0 of the servo valve type 25L is slightly larger than required no-load flow rate q_{0r}, Substituting $\eta_v q_0 = 0.295 \times 10^{-3} \, \mathrm{m}^3/\mathrm{s}$, $a\omega_b = 0.3 \, \mathrm{m/s}$ and $\phi = \pi/2$ into Eq. (4.214) yields the piston area A

$$A = A_\alpha = \frac{0.295 \times 10^{-3}}{\sqrt{3} \times 10^{-2} \times 30 \times \cos(\pi/4)} = 8.03 \times 10^{-4} \, \mathrm{m}^2 = 8.03 \, \mathrm{cm}^2. \tag{10}$$

The dimensions of the system element are finally given as

$$\left. \begin{array}{l} \text{Supply pressure:} \quad p_s = 7 \, \mathrm{MPa}, \quad \text{Piton area:} \quad A = 8 \, \mathrm{cm}^2 \\ \text{Servo valve:} \quad \text{Type 25L}(q_0 = 17.7 \, \mathrm{L}/\min \\ \qquad \qquad \text{for } p_s = 7 \, \mathrm{MPa} \quad \text{and} \quad I_R = 30 \, \mathrm{mA}) \end{array} \right\} \tag{11}$$

Substituting the rated maximum supply pressure $p_{sm} = 14$ MPa and the rated maximum flow rate $q_0 = 25$ L/min $= 4.17 \times 10^{-4}$ m^3/s into Eq. (4.202), the servo valve capacity factor CI_R is obtained as follows:

$$CI_R = 0.111 \times 10^{-6} \, \text{m}^3/(\text{Pa}^{1/2}\text{s}), \quad C = 0.371 \times 10^{-8} \, \text{m}^3/(\text{mA} \cdot \text{Pa}^{1/2}\text{s}).$$
$$(12)$$

Substituting $p_s = 7$ MPa, $A = 8 \, \text{cm}^2$, $CI_R = 0.111 \times 10^{-6} \, \text{m}^3/(\text{Pa}^{1/2}\text{s})$ and $\eta_v = \eta_c = 1$ into Eq. (4.201) yields the driving characteristic equation of the servo cylinder system:

$$V = (4.92 \times 10^{-3}) \times \sqrt{5600 - F} \, \text{m/s}. \qquad (13)$$

As shown in Fig. 4.70, the driving characteristic curve fully envelopes the load chart obtained recalling Eqs. (1), (2) and (3).

(3) *Discussions on control performance of a servo cylinder system*

It is worthwhile to take into account the control performances of the servo cylinder system even though the system designed in clause (2) satisfies the required power conditions. Considering the servo amplifier with the rated electric current $I_R = 30$ mA for rated maximum input voltage $V_R = 10$ V, the servo amplifier gain becomes $K_a = 3$ mA/V. The sensor gain K_t should be selected as $K_t = 10/(12 \times 10^{-3}) = 8.33 \times 10^2$ V/m so as to displace the piston by 12 mm for the maximum input voltage 10 V.

Considering only the inertia load due to the mass, the load pressure $p_{\ell 0}$ may be regarded as zero at the steady movement of the piston. Therefore, the gain constant $K_u = K_v K_m$ of the servo cylinder system can be

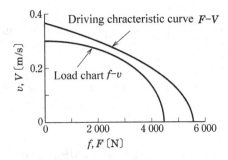

Fig. 4.70   Load chart and a driving characteristic curve.

expressed as the ratio of maximum piston speed $V = q_0/A$ to the rated electrical current $I_R = 30\,\text{mA}$. Recalling Eq. (4.202), the gain constant K_u is obtained as follows:

$$
\left.
\begin{aligned}
K_u &= K_v K_m = \frac{CI_R\sqrt{p_s}}{I_R A} = \frac{0.111 \times \sqrt{7} \times 10^{-3}}{30 \times (8 \times 10^{-4})} \\
&= 1.23 \times 10^{-2}\,\text{m}/(\text{s}\cdot\text{mA}) \\
K_0 &= K_t K_a K_v K_m = 3 \times (8.33 \times 10^2) \times (1.23 \times 10^{-2}) \\
&= 31.0 > 1/0.0334\,\text{rad/s}
\end{aligned}
\right\}.
\tag{14}
$$

Let us assume the following specifications of the servo system to estimate the dynamic characteristic parameters ω_n and ζ in Eq. (4.159).

The one side cylinder chamber volume including the pipeline volume between the piston and the servo valve: $V_0 = 32\,\text{cm}^3$ provided that piston is located at the cylinder center. Bulk modulus of the working oil is $\kappa = 1/\beta = 1 \times 10^3$ MPa. The density of the working oil is $\rho = 0.85 \times 10^3\,\text{kg/m}^3$. The ratio of under-lap clearance c to the maximum spool displacement x_m is $c/x_m = i_c/I_R = 0.07$ where i_c denotes the servo amplifier current corresponding to the under-lap clearance c, and $i_c = 2.1$ mA and $I_R = 30$ mA. Damping force and friction/spring force are regarded as zero.

Substituting pressure $p_{\ell 0} = 0$ into Eq. (4.128), the cylinder chamber pressures p_{10} and p_{20} are obtained as $p_{10} = p_{20} = p_s/2$ in the steady state. For cylinder chamber pressure $p_{10} = p_{20} = p_s/2$, the damping coefficient ζ in Eq. (4.160) is given as follows:

$$
\zeta = \sqrt{\frac{\kappa m}{2V_0}\frac{k_2}{A}}.
\tag{15}
$$

For response behaviors at $i_0 \approx 0$ in the infinitesimal input Δi, the parameter k_2 is given by substituting $i_0 = 0$ into Eq. (4.166). Recalling Eq. (15), the damping coefficient ζ is expressed as follows:

$$
\begin{aligned}
\zeta &= \sqrt{\frac{\kappa m}{2V_0 p_s}\frac{Ci_c}{A}} \\
&= \sqrt{\frac{(1 \times 10^9) \times 500}{2 \times (3.2 \times 10^{-5}) \times (7 \times 10^6)} \times \frac{(0.371 \times 10^{-8}) \times 2.1}{8 \times 10^{-4}}} = 0.33.
\end{aligned}
\tag{16}
$$

For response behaviors at $i_0 \approx 27$ mA in the infinitesimal input Δi, the parameter k_2 is obtained by substituting $i_0 = i_c + 27$ mA into Eq. (4.155).

Then the damping coefficient ζ is defined as

$$\zeta = \sqrt{\frac{\kappa m}{2V_0 p_s}} \frac{C(i_c + i_0)}{2A}$$

$$= \sqrt{\frac{(1 \times 10^9) \times 500}{2 \times (3.2 \times 10^{-5}) \times (7 \times 10^6)}} \times \frac{(0.371 \times 10^{-8}) \times 29.1}{2 \times (8 \times 10^{-4})} = 2.3. \quad (17)$$

Recalling Eq. (4.160), the undamped natural angular frequency ω_n is given as follows:

$$\omega_n = A\sqrt{\frac{2\kappa}{V_0 m}} = 8 \times 10^{-4} \times \sqrt{\frac{2 \times 10^9}{(3.2 \times 10^{-5}) \times 500}} = 283\,\text{rad/s}. \quad (18)$$

Therefore, the open-loop transfer functions $\overline{G}(s) = K_a K_t G_u(s)/s$ in Fig. 4.69 are denoted as follows:

$$i_0 = 0 \text{ mA}: \quad \overline{G}(s) = \frac{31 \times (8 \times 10^4)}{s(s^2 + 187s + 8 \times 10^4)}, \quad (19)$$

$$i_0 = 27 \text{ mA}: \quad \overline{G}(s) = \frac{31 \times (8 \times 10^4)}{s(s^2 + 1300s + 8 \times 10^4)}. \quad (20)$$

Since $\omega_n = 283\,\text{rad/s}$ satisfies the condition implicated by Eq. (4.225) for gain $K_0 = 31\,\text{rad/s}$, the gain crossover angular frequency ω_g is denoted as $\omega_g = K_0 = 31\,\text{rad/s}$ in accordance with Eq. (4.226).

Figure 4.71 shows the Bode diagrams of the open loop transfer function $\overline{G}(j\omega)$ for $i_0 = 0\,\text{mA}$ and $i_0 = 27\,\text{mA}$. In case of $i_0 = 0\,\text{mA}$, the Bode diagram reveals a phase margin $\phi_1 = 86°$ and gain crossover frequency $\omega_g = 31\,\text{rad/s}$. The result is satisfactory from the point of view of response characteristics and stability. On the other hand, in the case of $i_0 = 27\,\text{mA}$, the Bode diagram reveals a phase margin $\phi_1 = 64°$ and $\omega_g = 29\,\text{rad/s}$.

Since the control performance for $i_0 = 27\,\text{mA}$ is rather unsatisfactory with respect to the design specifications, the servo system should be improved by using compensation techniques. Figure 4.72 shows the indicial responses of control system corresponding to Eqs. (19) and (20).

(4) *Phase lead-lag compensation*

Using the phase lead-lag compensator, the control performance for $i_0 = 27\,\text{mA}$ can be improved so as to satisfy the design specifications i.e. $\phi_d = 80°$ and $\omega_{gd} = 32\,\text{rad/s}$. The transfer functions $G_{c1}(s)$ and $G_{c2}(s)$ in Eq. (4.245) are estimated by the following procedure.

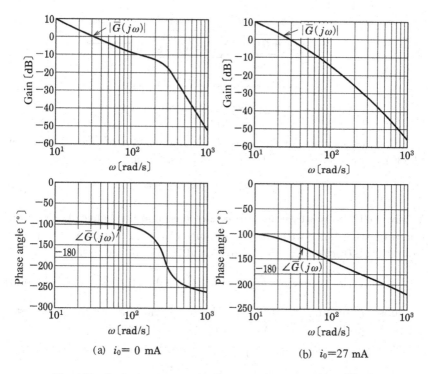

Fig. 4.71   Bode diagrams of the open loop transfer function $\overline{G}(s)$.

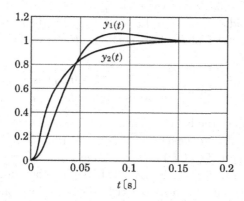

Fig. 4.72   Indicial response curves ($y_1(t)$: $i_0 = 0$ mA, $y_2(t)$: $i_0 = 27$ mA).

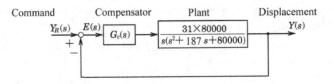

Fig. 4.73   Hydraulic servo system with a compensator.

First, let us consider the phase lead compensator of $G_{c1}(s)$. Equation (4.239) yields the phase angle ϕ_0 to compensate the phase margin:

$$\phi_0 = \phi_d - \phi_1 + \alpha = 80° - 64° + \alpha \qquad (21)$$

Where the factor α is estimated from the phase margin change of $\overline{G}(j\omega)$ produced in accordance with the transference of the gain crossover frequency from $\omega_g = 29\,\mathrm{rad/s}$ to $\omega_{gd} = 32\,\mathrm{rad/s}$, and then expressed as:

$$\alpha = \angle\overline{G}(j\omega_{gd}) - \angle\overline{G}(j\omega_g) \approx 4° \qquad (22)$$

By substituting Eq. (22) into Eq. (21), $\phi_m = \phi_0 = 20°$ is obtained. Rearranging Eq. (4.237), the factor a is represented as follows:

$$a = \frac{1 + \sin\phi_m}{1 - \sin\phi_m} = 2.04 \qquad (23)$$

The factor T may be estimated for the phase angle $\angle G_{c1}(j\omega)$ such that it should have the maximum value at $\omega_m = \omega_{gd} = 32\,\mathrm{rad/s}$. Substituting $\omega_m = 32\,\mathrm{rad/s}$ and $a = 2.04$ into Eq. (4.236) yields the factor $T = 0.0219\,\mathrm{s}$, and consequently, the phase lead compensator transfer function $G_{c1}(s)$ becomes:

$$G_{c1}(s) = \frac{1 + aTs}{1 + Ts} = \frac{1 + 0.0446\,s}{1 + 0.0219\,s} \qquad (24)$$

Next, let us consider the lag compensator of $G_{c2}(s)$. As shown in Fig. 4.74, the crossover angular frequency of $\overline{G}_1(j\omega) = G_{c1}(j\omega)\overline{G}(j\omega)$ is $42\,\mathrm{rad/s}$ and the logarithmic gain of $\overline{G}_1(j\omega)$ is $\Delta = 2\,\mathrm{dB}$ at $\omega_2 = \omega_{gd} = 32\,\mathrm{rad/s}$. Recalling Eqs. (4.243), (4.244) and Fig. 4.63, the lag compensator factors a and T in $G_{c2}(s)$ are estimated so as to satisfy the following conditions:

$$\left.\begin{array}{c} a = 10^{-\Delta/20} \\[2ex] T = \dfrac{10}{a\omega_2} \end{array}\right\} \qquad (25)$$

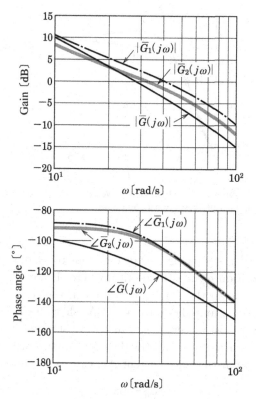

Fig. 4.74   Bode diagrams of $\overline{G}(s), \overline{G}_1(s)$ and $G_2(s)$.

From Eq. (25), factors $a = 0.794$ and $T = 0.394\,\text{s}$ are obtained. Therefore, the lag compensator transfer function $G_{c2}(s)$ becomes:

$$G_{c2}(s) = \frac{1 + aTs}{1 + Ts} = \frac{1 + 0.313\,s}{1 + 0.394\,s} \tag{26}$$

Finally, by using Eq. (4.245) the phase lead-lag compensator transfer function $G_c(s) = G_{c1}(s)G_{c2}(s)$ can be expressed:

$$G_c(s) = \frac{(1 + 0.0446\,s)(1 + 0.313\,s)}{(1 + 0.0219\,s)(1 + 0.394\,s)} \tag{27}$$

Figure 4.74 shows the Bode diagram of $\overline{G}_2(j\omega) = G_c(j\omega)\overline{G}(j\omega)$ which reveals the phase margin of $\phi_m = 80°$ and the crossover frequency of $\omega_2 = 32\,\text{rad/s}$. Figure 4.75 shows the results of indicial responses with and without the compensator for $i_0 = 27\,\text{mA}$.

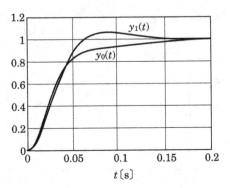

Fig. 4.75   Indicial responses of the hydraulic servo system with and without compensator ($y_0(t)$:without compensator, $y_1(t)$: with compensator).

(5) *Steady accuracy*

The servo cylinder system is a type-one control system. Therefore, the steady state position error is theoretically zero. However, steady position error creeps in due to the Coulomb friction in the servo valve or due to the drift in the servo amplifier output etc. As shown in Fig. 4.57, the practical steady state position error ε_p can be investigated by converting the miscellaneous disturbance to electrical current disturbance I_d.

The electrical disturbance current Δi_d can be expressed as:

$$\Delta i_d \leq 0.02 I_R \tag{28}$$

Equation (4.231) gives:

$$\varepsilon_p \leq \frac{\Delta i_d}{I_R} \frac{I_R}{K_t K_a} = 0.02 \times \frac{30}{(8.33 \times 10^2) \times 3} = 2.4 \times 10^{-4}\,\mathrm{m} = 0.24\,\mathrm{mm} \tag{29}$$

Problems

4.1 Derive the transfer function and the indicial response $y(t)$ provided that the relation between the input $x(t)$ and output $y(t)$ in the system is given by following equation.

$$\frac{d^2 y}{dt^2} + 4\frac{dy}{dt} + 5y = \frac{dx}{dt} + 3x.$$

4.2 Derive the inverse Laplace transformation of $y(t) = \mathcal{L}^{-1}[Y(s)]$.

$$Y(s) = \frac{s^2 + 2s + 5}{s(s+1)^2}.$$

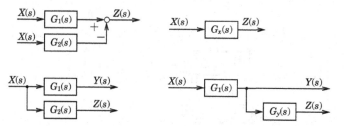

Fig. 4.76 Rearrangement in the partial diagram.

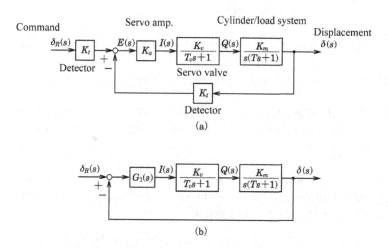

(a)

(b)

Fig. 4.77 Rearrangement in the block diagram.

4.3 In terms of the relationship between the input and output, the right side block diagram is equivalent to the left side block diagram in Fig. 4.76. Determine $G_x(s)$ and $G_y(s)$ using $G_1(s)$ and $G_2(s)$.

4.4 Determine the transfer function $G_1(s)$ and the closed loop transfer function $\delta(s)/\delta_R(s)$ in case when Fig. 4.77(a) is equivalent to Fig. 4.77(b).

4.5 The indicial response of a plant exhibits the control performances with the first peak time $t_p = 0.2\,$s, settling time $t_s = 1.0\,$s for error $\Delta = 5\%$ and steady state position $y(\infty) = 5$. Show the transfer function $G(s)$ of the plant provided that the plant performance may be regarded as a standard second order lag system.

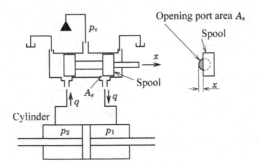

Fig. 4.78   Servo cylinder system.

4.6 Plot the Bode diagrams of transfer functions $G_1(s)$ and $G_2(s)$.

$$(1) \ G_1(s) = \frac{0.6s + 1}{0.2s + 1} \quad (2) \ G_2(s) = \frac{0.2s + 1}{0.6s + 1}.$$

4.7 Plot the Bode diagram of the transfer function $G(s) = 2/[s(2s + 1)]$, and then compare it with the Bode diagram approximated by the straight-lines.

4.8 Figure 4.78 illustrates the servo cylinder system controlled by changing the opening area A_s of the circular spool port with diameter d. The opening area A_s is given as

$$\text{For } \frac{x}{d} \le 0.5: \quad A_s = 1.7 \left(\frac{x}{d}\right)^{1.5} \left(1 - \frac{0.3x}{d}\right) \left(\frac{\pi d^2}{4}\right)$$

(1) Write the equation of the flow rate q into the cylinder through the spool port using the following notation: spool displacement x, flow coefficient C_d of the valve port with diameter d, differential pressure p_ℓ between the left and right cylinder chamber, supply pressure p_s and working oil density ρ.

(2) Derive the linearized equation of the flow rate change Δq when Δq is regarded as a small deviation from the equilibrium state at the differential pressure $p_\ell = p_{\ell 0}$, spool displacement $x = x_0$.

4.9 Answer the following questions, as in the [Example 4.2] given in Section 4.4.4, for the supply pressure $p_s = 10$ MPa.

(1) Estimate the piston area A and select the adequate servo valve type from options summarized in Table 4.10.

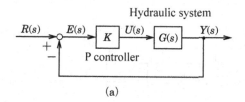

(a)

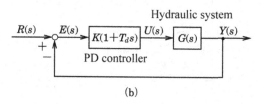

(b)

Fig. 4.79　Hydraulic systems with controllers.

(2) Derive the driving characteristic equation of a servo cylinder system.

4.10 For the hydraulic system shown in Fig. 4.79(a), the relation between input $u(t)$ and output $y(t)$ is given as

$$\frac{d^2y(t)}{dt^2} + 4\frac{dy(t)}{dt} = 4u(t),$$

where $R(s), E(s), U(s)$ and $Y(s)$, respectively denote the Laplace transformations of time variable functions $r(t), e(t), u(t)$ and $y(t)$.

(1) Derive the transfer function $G(s) = Y(s)/U(s)$ of the hydraulic system.

(2) Determine the error signal $E(s)$ using $R(s), G(s)$ and K for the system shown in Fig. 4.79(a).

(3) Find the range of K corresponding to a steady state velocity error $e_v(\infty) \le 0.1$ for the ramp input $r(t) = t$ as in the control system shown in Fig. 4.79(a).

(4) As shown in Fig. 4.79(b), a PD controller is used to improve the stability of the control system in Fig. 4.79(a). Firstly, determine the closed loop transfer function of the control system in Fig. 4.79(b). Secondly, estimate the derivative time T_D for $K = 16$ so as not to produce the oscillatory response $y(t)$.

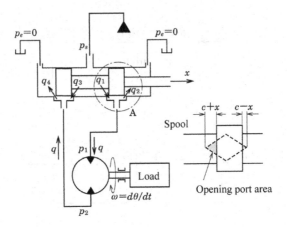

Fig. 4.80   Hydraulic motor control system.

4.11 Figure 4.80 shows a hydraulic motor system controlled by the servo valve with an under-lap clearance c. When the valve spool is located at distance $x = x_0 \leq c$ from the neutral position, it opens the spool port with a regular triangle area, its height $x_0 \pm c$ as shown in Fig. 4.80. Then the flow at the rate $q = q_0$ steadily enters the hydraulic motor chamber opening to the spool port. If the spool displacement slightly changes from $x = x_0$ to $x = x_0 + \Delta x(t)$, then the flow rate $q = q_0$ changes to $q = q_0 + \Delta q(t)$ and the differential pressure $p_\ell = p_{\ell 0}$ will simultaneously change to $p_\ell = p_{\ell 0} + \Delta p_\ell(t)$. Derive the linearized equation of the small flow rate change $\Delta q(t)$ as a function in the small variables $\Delta p_\ell(t)$ and $\Delta x(t)$. The supply pressure p_s and flow coefficient C_d of the spool port are assumed to be constants.

4.12 In the servo-cylinder system with the capacity factor $CI_R = 0.158 \times 10^{-6}\ \mathrm{Pa}^{-1/2}\,\mathrm{m}^3/\mathrm{s}$, the driving characteristics curve encloses a load chart joining with it at a point P ($F_t = 4500$ N, $V_t = 0.39$ m/s) and the maximum thrust is $F_m = 9000$ N. Assuming the both volumetric and thrust efficiency as one, calculate the maximum cylinder speed V_m, the supply pressure P_s to the servo valve, the cylinder area A, and the maximum power W_p.

4.13 The open-loop transfer function $G(s)$ in Fig. 4.81(a) is given as follows:

$$G(s) = \frac{32 \times 90000}{s(s^2 + 1400s + 90000)}.$$

(a) System without compensator

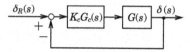

(b) System with compensator

Fig. 4.81 Control system with and without compensator.

As shown in Fig. 4.81(b), a series compensator with a transfer function $K_c G_c(s)$ is used to improve the control characteristics.

(1) Assuming $G_c(s) = 1$, estimate the adjusting parameter K_c to reduce the steady velocity error of the control system in Fig. 4.81(a) to one-third of its original value.

(2) Estimate factors a and T in the phase lag compensator parameter such that the stability and the response speed of the system should not deteriorate for the gain K_c obtained as the solution to the previous problem.

References

(1) Watton, J., *Fluid Power Systems* (1989), Prentice Hall, New Jersey.

(2) Esposito, A., *Fluid Power with Applications* (1994), Prentice Hall, New Jersey.

(3) Green, J. W., *Aircraft Hydraulic Systems* (1985), John Wiley and Sons, New Jersey.

(4) Ogata, K., *Modern Control Engineering* (1997), Prentice Hall, New Jersey.

(5) Merritt, H. E., *Hydraulic Control System* (1967), John Wiley and Sons, New Jersey.

(6) Shames, I. H., *Mechanics of Fluid* (1962), McGraw-Hill, New York.

(7) Sullivan, J. A., *Fluid Power-Theory and Applications* (1989), Prentice Hall, New Jersey.

(8) Blackburn, J. F., *Fluid Power Control* (1960), MIT Press, Massachusetts.

(9) Walters, R. B., *Hydraulic and Electro-hydraulic Control Systems* (1991), Elsevier Applied Science, Amsterdam.

(10) Nishiumi, T., Konami, S. and Uchino, T., An Application of the Identification Method Using Self-excited Oscillation to a Hydraulic Motor/load System, *Proc. of the 10th Bath International Fluid Power Workshop*, Bath (1997), pp. 381–395.

(11) Nishiumi, T. and Konami, S., A Design Technique of Hydraulic Servo Actuator Systems for Effective Power Transmission, *Proc. of the 4th JFPS International Symposium on Fluid Power*, Tokyo (1999), pp. 279–284.

(12) Konami, S., Nishiumi, T. and Hata, K., Identification of Linearized Electro-Hydraulic Servo-valve Dynamics by Analyzing Self-excited Oscillations, *Journal of the Japan Hydraulic Pneumatic Society*, Vol. 27, No. 4 (1997), pp. 143–149. (in Japanese).

(13) Nishiumi, T., Ichiyanagi, T., Katoh, H. and Konami, S., Real-time Parameter Estimation of Hydraulic Servo Actuator Systems Using Self-excited Oscillation Method, *Transactions of the Japan Fluid Power System Society*, Vol. 36, No. 1 (2005), pp. 1–7. (in Japanese).

(14) Ikebe, Y., Ikebe, J., Nakano, K. and Matushima, K., *Servo mechanism and Elements* (1965), Ohmsha Ltd., Tokyo (in Japanese).

(15) Ishihara, T., *Hydraulic Engineering* (1968), Asakura Publishing Co., Ltd., Tokyo (in Japanese).

(16) Ichikawa, T., *Hydraulics and Hydrodynamics* (1968), Asakura Publishing Co., Ltd., Tokyo (in Japanese).

(17) Ichikawa, T. and Hibi, A., *Hydraulic Engineering* (1979), Asakura Publishing Co., Ltd., Tokyo (in Japanese).

(18) Ichiryuu, K., *Electro Hydraulic Control* (1993), Nikkan Kogyo Shimbun, Ltd., Tokyo (in Japanese).

(19) Imaki, K., *Introduction to Hydraulic Engineering* (1991), Rikogakusha Publishing Co., Ltd., Tokyo (in Japanese).

(20) Koura, M. and Nakamura, T., *Expounder of Hydraulic Equipment* (1986), Seizando–Shoten Publishing Co., Ltd., Tokyo (in Japanese).

(21) Konami, S. and Degawa, T., *Introduction to Control Engineering* (1998), Gakukensha Publishing Co., Ltd., Tokyo (in Japanese).

(22) Satoh, T., *Design of Hydraulic Servo Systems* (1980), Taiga Publishing Co., Ltd., Tokyo (in Japanese).

(23) Tsuji, S., *Hydraulic Engineering* (1982), Nikkan Kogyo Shimbun, Ltd., Tokyo (in Japanese).

(24) Takenaka, T. and Urata, E., *Hydraulic Control* (1975), Maruzen Co., Ltd., Tokyo (in Japanese).

(25) Takenaka, T. and Urata, E., *Hydraulic Engineering* (1970), Yokendo Co., Ltd., Tokyo (in Japanese).

(26) Yamaguchi, T. and Tanaka, H., *Oil Hydraulics and Pneumatics* (1986), Corona Publishing Co., Ltd., Tokyo (in Japanese).

(27) Research Group in Dakin Ind. Ltd., Hydraulic Machinery Vol. 1 and Vol. 2 (1974), Japan Machinist-Sha Co., Ltd., Tokyo (in Japanese).

(28) Research Group in Nachi Fujikoshi Corp., Elucidation of Oil Hydraulics Application Edition (1984), Japan Machinist-Sha Co., Ltd. (in Japanese).

(29) Research Group in Nachi Fujikoshi Corp., Elucidation of Oil Hydraulics-Circuit and Data Edition (1989) Japan Machinist-Sha Co., Ltd. (in Japanese).

(30) Japan Fluid Power System Society, *Handbook of Oil Hydraulics and Pneumatics* (1989), Ohmsha Ltd., Tokyo (in Japanese).

Appendix

Table A.1　Graphic symbols for hydraulic diagrams.

(1) Symbol elements and graphic symbols of mechanical elements.

Symbols	Annotation	Symbols	Annotation
———	Pipe line, Electro signal line	⊔	Open type tank
---------	Pilot line, Drain line, Transient position	⬭	Tight type oil tank, Accumulator
-·-·-·-·-	Surrounding line sigifies grouping of elements	▶	Pressure oil: Vertex denotes flow direction through valve
$L/5$ / L	Mechanical unions such as shaft, lever, piston rod	↙ ↑	Flow direction through valve, Direction of straight motion
◯	Circle with diameter $d=L$, Energy transformer	↰ ↳	Direction of shaft rotation, Rotational motion
◯	Circle with diameter $d=3L/4$, Measuring meter	↗	Variable function on displacement, Spring force, Throttle area etc., being added with respective symbols
○	Circle with diameter $d=L/4{\sim}L/3$, Check valve element etc.	⊥	Closed pipe line and port
D	Semicircle with diameter $d=L$, Limited rotary motor/pump	M	Spring
		⋈	Throttle
☐	Square with length L, Prime mover excluding electric motor	—◡—	Hose
		⊥	Joint pipe
		+ ⤴	Crossover pipe lines
◇	Rhombus with length L Fluid conditioner	☐	Square with side length ($L/2$), Cushion in cylinder, Weight in accumulator

(*Continued*)

303

Table A.1   (*Continued*)

(2) Symbols of hydraulic instruments.

Symbols	Annotation	Symbols	Annotation
	Hydraulic source		Prime mover excluding electric motor
	Electric motor		Left: hydraulic pump Right: hydraulic motor
	Fixed displacement pump with one directional flow and revolution by electric motor drive		Variable displacement pump with two directional flow and one directional revolution
	Fixed displacement motor with two directional flow and revolution		Variable displacement pump with pressure compensation
	Fixed displacement pump /motor with one directional flow and revolution		Rotary actuator
	Double acting single rod cylinder		Double acting double rod cylinder without and with cushion
	Single acting cylinder wtih return spring		Telescope type cylinder single and double acting
	Accumulator with and without initial gas pressure		One directional flow control with two ports and two position
	Four direction flow control valve		Two directional flow control valve
	Check valve with and without spring		Servo valve
	Pilot operated check valve with and without spring		Four directional flow control valve
	Relief valve		Shuttle valve
	Pilot operated relief valve by solenoid valve		Pilot operated relief vlave
	Sequence valve with outer drain and outer pilot		Reducing valve
	Unload valve without outer drain		Proportional solenoid control reducing valve
			Counter balance valve

(*Continued*)

Table A.1   (*Continued*)

Symbols	Annotation	Symbols	Annotation
	Variable throttle		Stop valve
	Flow control valve with check valve		Deceleration valve
	Flow control valve with pressure compensation and temperature compensation		Flow divider
	Pressurized oil tank		Opened oil tank with pipe under and upper oil surface
	Oil filter Oil filter with silting meter		Heat exchager
			Heater
	Pressure indicator without pick up sensor		Heat regulator
	Pressure gauge		Pressure difference meter
	Oil level meter		Thermometer
	Galvanometer		Flow meter
	Revolution speed meter		Torque meter
	Pressure switch		Limit switch

Table A.2   Prefixes in SI unit system and factor.

Symbol	Prefix	Factor	Symbol	Prefix	Factor
G	giga-	10^9	m	milli-	10^{-3}
M	mega-	10^6	μ	micro-	10^{-6}
k	kilo-	10^3	n	nano-	10^{-9}
c	centi-	10^{-2}	p	pico-	10^{-12}

Table A.3 SI unit systems and other unit systems.

Quantity	SI unit systems	Other unit systems	Conversion factors
Force	newton [N] $1\,[\mathrm{N}] = 1\,[\mathrm{kg \cdot m/s^2}]$	weight kilogram [kgf] $1\,[\mathrm{kgf}] = 1\,[\mathrm{kg}] \times 9.81\,[\mathrm{m/s^2}]$	$1\,[\mathrm{kgf}] = 9.81\,[\mathrm{N}]$
Mass	kilogram [kg]	$[\mathrm{kgf/(m/s^2)}]$	$1\,[\mathrm{kg}] = 0.102\,[\mathrm{kgf/(m/s^2)}]$
Work (Energy)	joule [J] $1\,[\mathrm{J}] = 1\,[\mathrm{Nm}]$	calorie [cal] $1\,[\mathrm{kcal}] = 427\,[\mathrm{kgf \cdot m}]$	$1\,[\mathrm{kgf \cdot m}] = 9.81\,[\mathrm{J}]$ $1\,[\mathrm{cal}] = 4.19\,[\mathrm{J}]$
Power	watt [W] $1\,[\mathrm{W}] = 1\,[\mathrm{J/s}]$	French Horse Power [PS] $1\,[\mathrm{PS}] = 75\,[\mathrm{kgf \cdot m/s}]$	$1\,[\mathrm{kgf \cdot m/s}] = 9.81\,[\mathrm{W}]$ $1\,[\mathrm{PS}] = 735.5\,[\mathrm{W}]$
Pressure	pascal [Pa] $1\,[\mathrm{Pa}] = 1\,[\mathrm{N/m^2}]$	$\mathrm{kgf/cm^2}$, atm, bar $1\,[\mathrm{atm}] = 1.033\,[\mathrm{kgf/cm^2}]$	$1\,[\mathrm{kgf/cm^2}] = 9.81 \times 10^4\,[\mathrm{Pa}]$ $1\,[\mathrm{atm}] = 1.0133 \times 10^5\,[\mathrm{Pa}]$ $1\,[\mathrm{bar}] = 10^5\,[\mathrm{Pa}]$
Flow rate	$[\mathrm{m^3/s}]$	liter per minute [L/min]	$1\,[\mathrm{L/min}] = 1.67 \times 10^{-5}\,[\mathrm{m^3/s}]$
Angular velocity	[rad/s]	revolutions per minute [rpm]	$1\,[\mathrm{rpm}] = 1\,[\mathrm{min^{-1}}] = 0.1047\,[\mathrm{rad/s}]$
Viscosity	[Pa·s]	poise [P], centi-poise [cP] $1\,[\mathrm{cP}] = 1 \times 10^{-2}\,[\mathrm{P}]$	$1\,[\mathrm{P}] = 0.1\,[\mathrm{Pa \cdot s}]$
Kinematic viscosity	$[\mathrm{m^2/s}]$	stokes [St], centi-stokes [cSt] $1\,[\mathrm{St}] = 1\,[\mathrm{cm^2/s}]$ $1\,[\mathrm{cSt}] = 1 \times 10^{-2}\,[\mathrm{St}]$	$1\,[\mathrm{St}] = 1 \times 10^{-4}\,[\mathrm{m^2/s}]$

Answers to Problems

Chapter 2

2.1 10.0 N

2.2 Volume change: $-3.92 \times 10^{-4}\,\text{m}^3$

2.3 $\kappa = 0.91 \times 10^3\,\text{MPa}$

2.4 5.74°C, $W = 3.33\,\text{kW}$

2.5 4045 N

2.6 $f_x = \rho Q |u| \sin\theta$, $\quad Q_1 = \dfrac{Q(1 - \cos\theta)}{2}$, $\quad Q_2 = \dfrac{Q(1 + \cos\theta)}{2}$

2.7 (1) $\dfrac{\Delta p_1}{\Delta p_2} = n^{-1.75}$ (2) $\dfrac{\Delta p_1}{\Delta p_2} = \dfrac{1}{n}$

2.8 $\dfrac{Q_2}{Q_1} = \left(\dfrac{d_2}{d_1}\right)^{19/7} \left(\dfrac{\ell_1}{\ell_2}\right)^{4/7}$

2.9 $\dfrac{\Delta p_2}{\Delta p_1} = 0.663$

2.10 $Q = 4.19\,\text{cm}^3/\text{s}$, $\quad Q/Q_0 = 1 + 1.5(e/h)^2$

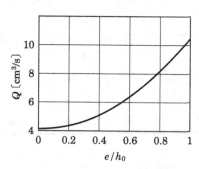

Fig. 1 Leakage flow rate through bore-spool clearance.

307

2.11 $p_s = 13.1\,\mathrm{MPa}, \quad h = 68.4\,\mu\mathrm{m}$

2.12 (1) $p = \dfrac{3\mu v r_0^2}{h^3}\left[1 - \left(\dfrac{r}{r_0}\right)^2\right]$ (2) $F = \dfrac{3\pi\mu v r_0^4}{2h^3}$

Chapter 3

3.1 (1) $2\pi D_p = 29.6\,\mathrm{cm^3/rev}$ (2) $W_i = 28.6\,\mathrm{kW}$ (3) $T_a = 152\,\mathrm{N\cdot m}$

3.2 $T_a = 35.2\,\mathrm{N\cdot m}$

3.3 (1) $T_m = 989\,\mathrm{N\cdot m}$ (2) $2\pi D_m = 6.22\times 10^{-4}\,\mathrm{m^3/rev}, \quad \alpha = 3.5°$
     (3) $\omega_m = 26.4\,\mathrm{rad/s}$

3.4 $\eta_{pm} = 0.9, \quad \dfrac{p}{\mu\omega} = 4.85\times 10^4$

3.5 $\omega = 5\,\mathrm{rad/s}$

3.6 Load torque $T = -J\theta_a\omega_a\sin\omega_a t$

     Load power $W = \dfrac{J\theta_a^2\omega_a^3\sin 2\omega_a t}{2}$

     $r_0 = 64.9\,\mathrm{mm}, \quad W_{\max}/\eta_m = 4.41\,\mathrm{kW}$

3.7 (1) $V_{m1} = \dfrac{V_0(n-1)}{n\cdot\ln n}$,   (2) $V_{m2} = V_0\dfrac{\ln n}{n-1}$
     (3)

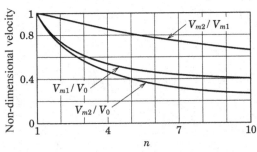

Fig. 2   Mean non-dimensional velocity for the damping rate.

3.8 $\Delta p = 15.4\,\mathrm{MPa}, \quad F = -233.6\,\mathrm{N}$ (Force opening the port is defined
     as positive)

3.9 $x = \dfrac{\pi d^2\Delta p - 4\delta k}{4(\pi C_d d\Delta p\sin 2\alpha + k)} = 1.11\,\mathrm{mm}$

3.10 $\dfrac{\Delta p}{p_s} = 1 - \dfrac{1 - [(2C_d A_x \cos\alpha)/S]}{1 + (x/x_0)} = 0.419$

3.11 (1) Static balancing force equation:

$$A(p_s - p_c) - (\rho Q^2/a_0) = (10 + n)kL$$

Flow rate passing through the fixed orifice:

$$Q = C_d a_0 \sqrt{\dfrac{2(p_s - p_c)}{\rho}}$$

Flow rate passing through the variable orifice:

$$Q = C_{dv} w(L - nL) \sqrt{\dfrac{2(p_c - p_L)}{\rho}}$$

(2) $\dfrac{Q_2}{Q_1} = \sqrt{\dfrac{10 + n_2}{10 + n_1}} = 1.024$

$$\dfrac{\Delta p_2}{\Delta p_1} = \dfrac{10 + n_2}{10 + n_1} \times \dfrac{1 + \left(\frac{a_0}{wL}\right)^2 \frac{1}{(1-n_2)^2}}{1 + \left(\frac{a_0}{wL}\right)^2 \frac{1}{(1-n_1)^2}} = 6.42$$

3.12 $\delta = \dfrac{\rho Q_0^2 A}{2k(C_d a)^2}\left(1 - \dfrac{2C_d a}{A}\right)$

3.13

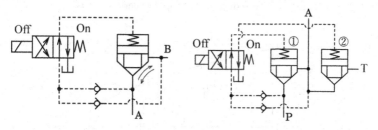

Fig. 3   Hydraulic logic circuits.

3.14 Recalling Fig. 3.102, Polytropic exponent of nitrogen gas is $n = 1.9$ for the mean pressure $p_s = (12 + 20)/2 = 16\,\text{MPa}$. Recalling Eq. (3.127) and $\eta = 0.95$,

$$\Delta V = \dfrac{14 \times 0.95 \times \left[\left(\frac{20}{15}\right)^{0.5263} - 1\right]}{\left(\frac{15}{12}\right)\left(\frac{20}{15}\right)^{1/1.9}} = 1.5\,\text{L}.$$

Chapter 4

4.1 $\dfrac{Y(s)}{X(s)} = \dfrac{s+3}{s^2+4s+5}$,   For $x(t) = u(t) : y(t) = \sqrt{2}e^{-2t}\sin[t+(\pi/4)]$

4.2 $t \geq 0 : y(t) = -4e^{-t}(t+1) + 5$, $t < 0 : y(t) = 0$

4.3 $G_x(s) = G_1(s) - G_2(s)$, $G_y(s) = G_2(s)/G_1(s)$

4.4 $G_x(s) = K_a K_t$, $\dfrac{\delta(s)}{\delta_R} = \dfrac{K_a K_t K_v K_m}{s(T_v s + 1)(Ts + 1) + K_a K_t K_v K_m}$

4.5 $G(s) = \dfrac{1280}{s^2 + 6s + 256}$

4.6

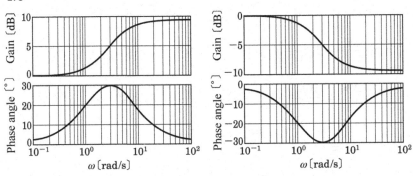

Fig. 4   Bode diagrams of the transfer functions.

4.7

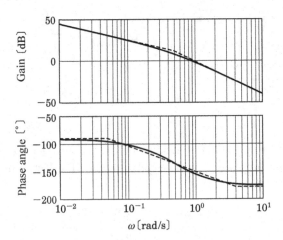

Fig. 5   Bode diagram compared with the straight approximations.

4.8 (1) $q = 1.7C_d \left(\dfrac{x}{d}\right)^{1.5} \left(1 - \dfrac{0.3x}{d}\right) \dfrac{\pi d^2}{4} \sqrt{\dfrac{p_s - p_\ell}{\rho}}$

(2) $\Delta q = k_1 \Delta x + \Delta k_2 \Delta p_\ell$

$$k_1 = \frac{2.55\pi C_d d}{4} \sqrt{\frac{x_0}{d}} \left(1 - \frac{0.5x_0}{d}\right) \sqrt{\frac{p_s - p_{\ell 0}}{\rho}}$$

$$k_2 = \frac{-1.7\pi C_d d^2}{8} \left(\frac{x_0}{d}\right)^{1.5} \left(1 - \frac{0.3x_0}{d}\right) \frac{1}{\sqrt{\rho(p_s - p_{\ell 0})}}$$

4.9 (1) $A = 5.6$ cm^2 Servo valve: Type15L in Table 4.10

(2) $V = 5.04 \times 10^{-3} \times \sqrt{5600 - F}$

4.10 (1) $\dfrac{Y(s)}{U(s)} = \dfrac{4}{s(s+4)}$ (2) $E(s) = \dfrac{R(s)}{1 + KG(s)}$ (3) $K \geq 10$

(4) (a) $\dfrac{Y(s)}{R(s)} = \dfrac{4K(1 + T_d s)}{s^2 + 4(1 + KT_d)s + 4K}$ (b) $T_d > 0.118$

4.11 $|x| \leq c : q = \dfrac{C_d(c+x)^2}{\sqrt{3\rho}} \sqrt{p_s - p_\ell} - \dfrac{C_d(c-x)^2}{\sqrt{3\rho}} \sqrt{p_s + p_\ell}$

$$\Delta q = k_1 \Delta x + k_2 \Delta p_\ell$$

$$k_1 = \frac{2C_d}{\sqrt{3\rho}} \left[(x_0 + c)\sqrt{p_s - p_{\ell 0}} - (x_0 - c)\sqrt{p_s + p_{\ell 0}}\right]$$

$$k_2 = -\frac{C_d}{2\sqrt{3\rho}} \left[\frac{(x_0 + c)^2}{\sqrt{p_s - p_{\ell 0}}} + \frac{(x - c)^2}{\sqrt{p_s + p_{\ell 0}}}\right]$$

4.12 $V_m = 0.552\,\text{m/s}$, $P_s = 9.96\,\text{MPa}$, $A = 9.04\,\text{cm}^2$, $W_p = 1.91\,\text{kW}$

4.13 (1) $K_c = 3$,   (2) $G_c = \dfrac{1 + 0.313s}{1 + 0.94s}$

Index

Printed in the United States
By Bookmasters